Investigating Being in Organizations and Leadership

Kim Malmbak Meltofte Møller
Michael Fast

Investigating Being in Organizations and Leadership

A Phenomenological Alternative

palgrave
macmillan

Kim Malmbak Meltofte Møller
Aalborg University Business School
Aalborg, Denmark

Michael Fast
University College Northern Denmark
UCN Technology & Business
Aalborg, Denmark

ISBN 978-3-030-58137-4 ISBN 978-3-030-58138-1 (eBook)
https://doi.org/10.1007/978-3-030-58138-1

This Palgrave Pivot imprint is published by the registered company Springer Nature Switzerland AG.
The registered company address is: Gewerbestrasse 11, 6330 Cham, Switzerland

Preface

This book is the result of decades of research experience in organizational life and human existence. It is the result of a desire to understand, to change things and make a difference. To shed light on our existence and on social structures regarding organizations and leadership in organizations that is currently locking us into certain perspectives regarding the future and limiting us in our thinking and acting.

Through the existential discussions in this book it is shown that any perception is always a result of emotional, historical and ethical considerations, which entails that all considerations, argumentations and scientific endeavors are always in relation to a context, position or individual. By accepting this notion, any theoretical description must take into account its relation to society and other relevant contexts. As such, theories and any related statements cannot be said to be neutral, objective, or in any other manner detached from their context and values therein, since they are a part of our perception and of the discourses in society.

The book is a possible alternative to what has been for decades the dominant business economic trends and perspectives on organizational and management research. With the current need to think and act in relation to environmental, political and societal change and responsibilities, we hope that this book can serve as one of the starting points for the upcoming (r)evolution within organizational research and practice.

Aalborg, Denmark

Kim Malmbak Meltofte Møller
Michael Fast

Contents

CHAPTER 1

Introduction to the Philosophical Investigation of Being in Organizations

Abstract In this chapter the relevance of an ontological discussion about organizations is established along with a brief description of the sources of philosophical inspiration for the book. The structure of the book is presented as it examines some of the perspectives on organizational life and existence as a human being.

Keywords Phenomenology • Lifeworld • Essence • Organizational assumptions

This book looks at life in organizations—the being in organizations. Most of us take this for granted, that to be in an organization is just something we do, as we work and live our everyday lives within them occupied with our assignments. We are engaged in many organizations during our lives; some of them are more important than others to us, and some are difficult to be in and to work in. No matter what, all of them will entail creating experiences that are important to our understanding of them, to the understanding of ourselves and to the life we live in them, as well as of the experiences of the actual creation of those organizations. It is this discussion of this being and creation of organizations that is the focus of this book, as organizations are nothing in themselves. Organizations can only exist as contexts, ideas and manifestations of human beings and their lived lives.

K. M. M. Møller, M. Fast, *Investigating Being in Organizations and Leadership*, https://doi.org/10.1007/978-3-030-58138-1_1

From a phenomenological perspective, we will discuss this life and being in organizations through some of the thoughts and central themes which are essential in developing a perspective of human beings in organizations. This will include reflections on existence and life in organizations, as the focus is on being in organizations and understanding the process of the human beings defining and acting in the everyday life of organizations. To understand organizations is to understand the essence of them and what it is that is the foundation—the human being, and the becoming of the organization as a social phenomenon. The focus of this is the discussion of the subject and how the subjects interactions and understandings constitute the organization as social, and how to understand this becoming of the organizational Lifeworld. The organizational world of life is what we daily live, experience, talk about and take for granted in all our activities, and what we see as natural in so many ways. At the same time, humans' approach to the world is naive, as in our natural attitude we are ignorant of the conditions for our organizational existence and the possibilities that are in this existence. In our daily experiences, humans naturally and naively take for granted the whole of reality, as a substance existing in itself and unconscious of itself, and thus also of the role that the subject itself plays in the experience of the world. On the other hand, this world of human beings and organizations is at the same time full of contradictions and the process of the world is dialectical. The organizational world appears as complex and something that exists in itself and outside the subject.

The discussion of being in organization takes as its point of departure an existential phenomenological reflection on organization and management, and from this examines some of the important questions and themes: the question of how to understand organizations and management, how to understand the human being in organizations, the questions of how to understand the development of organizations and how the organization becomes a social arena, and in all of this how to understand the human being as a moral being and the question of ethics in organizations. All of these themes and questions of organizations must in some way be a part of the discussions and development in any organization. They are also something that must be stressed upon, and have an influence on political discourses and a collective consciousness in society.

There is a current need to discuss alternatives to mainstream theories and the positivistic and rationalistic traditions that have been dominating the science of business and economics. This need arises from the question of seeing human beings as human beings, and not figures or functions in a perspective on organizations. Organizational life, we must suppose, is more than that. Situations and problems are always from a position and from a perspective, and as Freidrich Nietzsche (1882) once stated, "There are no facts, only interpretations". If this is true, the question arises: What are interpretations and what does this mean for understanding organizations and human beings in them? There is no doubt that the differences between positivism or rationalism and phenomenology are much more than a debate about the facts of organizations. It is a fundamental ontological discussion of what organizational reality is and what is life in organizations.

The discussion of the human being and existence is a reflection of how being and becoming is an aspect of consciousness, and of the dialectical process of thinking and acting in everyday life. An important issue in being is the dialectics of everyday life, and understanding the movements and changes in the being of organizations.

The discussion of organizational being is also a critique of mainstream organization theory, both in relation to the ontology of organizations as founded in subject-subjects and as constituted in everyday life, and in the light of a critical discourse on organizations. In some ways, this attempt to think alternatively from the perspective of organizations partly stems from Berger and Luckmann's (1966) discussion of the social construction of reality in connection to the construction of society and the processes involved. Their perspective was very much inspired by phenomenology and especially Alfred Schutz's thinking. It can also be seen in Weick's (1969) discussion of the social psychology of organizing and of enactment, and his critique of system theory in thinking organizations. His perspective is a processual understanding of organizing and as such inspired as well of phenomenology, both in relation to Berger and Luckmann and of Schutz. The alternative can also be seen in Silverman's (1970) discussion of an action frame of organizations, as he focuses upon people's actions in organizational settings and not on structures.

Much of this and other work from the mid-1960s and 1970s was critical to positivism and rationalism, and took as its point of departure the traditions of phenomenology, hermeneutics and critical theory. From this era, a discussion in social science and business economics continued as a new perspective, with a focus on social construction, trying to understand organizations from a perspective other than rationalism or functionalism. But much of this discussion omitted the philosophical roots of phenomenology and a philosophical investigation into ontological ideas, content and arguments. The discussion will go back to this and a phenomenological reflection on being, as well as the original ontological ideas of the human being and the becoming of the social. There is a need to see science as grounded in philosophy, as any understanding stems from a perspective of ontology and epistemology, and methodology is only the consequence of that perspective.

The tradition in our perspective is the original discussion of broader phenomenology, and will discuss aspects of Kant, Husserl, Heidegger, Schutz, Merleau-Ponty, Gadamer, Mead, Buber, Sartre, Løgstrup and others linked to an ontological discussion of understanding being and Lifeworld, and social becoming. Central to this is understanding the human being as a bodily being, as well as cognition, consciousness and acting in the world. To understand, act and think in the world there is a need for a perspective that involves the ontological reasons and arguments of the human being as subject to the becoming of the human being as social. Organizational life as a social involvement of all the aspects of others and the idea that humans are human with others sees the human being as ethical and moral, as without these attributes there could not be any social becoming.

The human being does not act alone in the world. We are situated, confronted with the other, and in this our actions have consequences for the organization and for ourselves. Organizations are, at the same time, both stable and volatile. On one hand we talk about organizations as something concrete, as something in itself. On the other hand actions and interactions have consequences, that is, they are interpreted in the attempt to make sense of the organization and events, and create something that is both in common and also creates common sense for people and people's understanding of each other. Thus the organization is both individual and at the same time a social phenomenon.

Our discussion of and interest in philosophy and the philosophy of science and reflections of the human being in organizations goes far back in time.

Michael, with his background in organizational sociology and philosophy of science, has been working on a critique of mainstream science in business economics and economics for many years. Some of his major works have been in cooperation with Dr. Woodrow W. Clark II, especially on the book *Qualitative Economics – The Science of Economics* (2019), as well as his discussion of being an entrepreneur (2018), and parts of those discussions can be seen in this book.

Kim has been focused for years on the existence of human beings and the whole discussion of a theory of cognition. He has also been focused on developing a phenomenology of being and the social in various discussions (2016, 2017 and 2017).

The Logic and Building of the Book

Chapter 2. The Individuality of Experience in Organizational Life: Reality is always seen, interpreted, understood and taken for granted from a certain point of belief and tradition. The origin of this and of all experiences of reality is the human being. This chapter discusses this along with a phenomenological ontology of an understanding of the human being, and the tradition and arguments for the world as a social and individual phenomenon. It highlights some of the important concepts of human beings used later as central to the discussion of humans and organizations, for example consciousness, intentionality, understanding, self-awareness and critique. The focus is on building an ontology and epistemology of understanding the essences of being and being together in organizations. It is a discussion of cognition and understanding of humanity and humanity's being in the world. Reality does not in itself have meaning. The meaning is connected to the thinking and acting human being. In turn, that thinking is a condition for existence and being. It is a condition for anything we can name a social context as the organization.

Chapter 3. The I and the Becoming of Social Being: One of the primary challenges inherent in the aspiration to describe humans as a general concept is that initially it assumes that the writer can somehow attach him/herself from the unique experiences and emotions of being a specific human being. Different elements can theoretically be assumed: the a priori elements of experiencing something, the process of experiencing something, the result of experiencing something. These three distinctions rely on the assumptions that everything has something prior to it that enables the existence of it, that everything has a time and place, and lastly that everything humans can experience exists. Our perception captures the result of things coming into our experience, but not the process. Moreover, this notion of things happening to us, which is commonly defined as something we do, an action we perform, will be highly relevant later. For now the question remains: What is a human being and how can such a thing be described with any sense of validity? What about the sentences I am writing now? Are they happening to me? Or are they performed and planned by me? It all depends on the definition of "me". The beginning to an answer is obviously that "me" or humans has to be defined as something which can both be happening and performing. In other words, we can carry out actions which are not consciously intended but rather unconsciously occurring and we are obviously able to carry out actions which have been intended prior to their execution. Furthermore, it is necessary to include both consciousness and unconsciousness in the definition of "me". In other words, I am more than I think.

At the same time, humans or organizational members can be misinterpreted as solely the product of social interactions, of cultural development or, as some state, tabula rasa. Given the argumentation so far, it is clear that the individual is to a large extent the product of social interaction—we learn from our current contexts. The individual constitutes society as genuinely as society constitutes the individual. Society can only exist when we have minds and selves, since all its most characteristic features presuppose the possession of minds and selves by its individual members, but its members would not possess minds and selves if these had not arisen within or emerged out of the human social process (see Mead 1962). So the questions are how do people meet and how do they construct and share meanings and a common understanding of each other and the organizational reality?

Chapter 4. Organizational Dialectics and the Becoming of the Organization: In history of philosophy, the question of how to understand development and changes in nature, history, society, people, and organizations have been central. There are situations and historical events that matters, but in which way and how can we connect those to social context and people changes. In phenomenology, the concept of dialectics is central in understanding development and is connected to the human being and social reality in different ways. Discussions of development and dialectics have taking place even before phenomenology was named as the ancient Greek philosophers discussed dialectic in different ways. Dialectic has been discussed in relation to cognition and knowledge, to method, to investigation of something other than the subject and cogito, and on conversation and dialogue. It is both discussions on how to understand and create knowledge (epistemology) and theorizing about the structure of the world (ontology). A world seen as something we are interwoven, with contradictions and dynamic before we observe it. So changes can be seen in the subject and in the social, and the development must be something that are in the subjects as well as in the social and what we meet in the world.

Chapter 5. Leadership as the Direction of Being. Throughout this book the analysis of being will provide a general description of what it means to be and to be an individual in an organization. Regarding popularity in literature, none are more so than those at the top of the organizational hierarchy, often referred to as managers or leaders. This chapter will distinguish between managers and leaders due to the qualitative difference between the characteristics of the two roles. The two concepts can be defined differently but since the goal is not an etymological discussion, a brief definition of the two concepts can be presented: a manager organizes and distributes resources, typically within firms, whereas a leader is highly significant in setting the cultural and/or practical direction of a group of people. Leadership is missing in organizations today and is substituted by managers who without explicit moral aspirations such as those found in leadership can without analysis accept and support ethically and culturally problematic assumptions regarding the human being.

Chapter 6. Existential Morality: All moral considerations stem from one of two narratives. The first is that morality originates from dilemmas in specific situations which encompass individual experience. Ethics is a choice, albeit not necessarily one we have the knowledge or competence to accurately act in accordance with. The second narrative claims that ethics is obedience to a set of moral principles or laws for which any contradictory actions are punished. We will for our discussions here disregard philosophies which are in line with the second narrative due to obvious inescapable religious ties. The first narrative, however, that morality is a choice in a specific situation which depends on knowledge and competences, does seem to indicate that ethics are a matter of the capacities of human beings. We further argue that this link should not only indicate the extent of morality but also its origin. Morality and ethical considerations are experiences for us precisely because we are human beings and the complexity of our existential functions creates the ability to imagine different outcomes and choices, and therefore we are existentially bound to navigating a sea of ethical potentials and corresponding intentions. We claim that the origins of morality are not intended to disregard the historical development of moral philosophy, but rather to point out that the very same historical development of moral philosophy is entirely situated upon human functions and therefore has to take into account that very same existential foundation. This entails that the origin, experiences and principles of morality can potentially be extrapolated from the existential functions of humans (to the extent of our incomplete knowledge of the workings of human beings).

Chapter 7. A Commentary on Sustainability and Organizational Life: This is a summarizing discussion regarding the existential functions of humans along with their implications for certain social structures. The purpose of the summary, beyond summarization itself, is to describe in broader terms the consequences of the analysis in this book and how it can be related to potential ideals for future organizational research and a more ethical organizational practice.

References

Berger, P L & Luckmann, T: "The Social Construction of Reality – a treatise in the sociology of knowledge." Doubleday & Company, New York, 1966.

Buber, M.: "I and Thou". Continuum, London. 2. Edition. Translated by Ronald Gregor Smith, 1958.

Clark, Woodrow W II & Fast, Michael: "Qualitative Economics: The Science of Economics." Springer International Publishing, October, 2019.

Fast, Michael: "Philosophical Perspective on Entrepreneurship." in R V Turcan & N M Fraser "The Palgrave Handbook of Multidisciplinary Perspectives on Entrepreneurship" Palgrave, 2018.

Gadamer, H-G: "Truth and Method". Sheed & Ward, London, 1993 (1975).

Heidegger, M.: "Being and time". Blackwell, Oxford, 1992 (1927).

Husserl, E.: "Ideas". Macmillan, New York, 1962.

Kant, I.: "Critique of pure reason" (*"Kritiken der reinen Vernunft" (1781/1787)*). Macmillan, Hong Kong, 1929 (1787).

Løgstrup, K.E. (Knud Ejler): "The Ethical Demand". Notre Dame: University of Notre Dame Press, 1997.

Mead, G H.: "Mind, Self, & Society – from the standpoint of a Social Behaviorist". The University of Chicago Press, Chicago, 1962 (1934).

Merleau-Ponty, M.: "The Phenomenology of Perception". Routledge & Kegan Paul, London, 1994 (1962).

Møller, Kim Malmbak Meltofte: "Cognition and organizational praxis: to wonder on the human being." (only in Danish Erkendelse og organisatorisk praksis: At undre sig over mennesket). Ph.D. thesis, Aalborg University, 2016.

Møller, Kim Malmbak Meltofte: "Experiencing a painting An interdisciplinary discussion regarding epistemology and experiencing." Academic Quarter, vol 16, Autumn 2017.

Møller, Kim Malmbak Meltofte & Fast, Michael: "Learning and Cognition – The interplay between the Subject and the Group: Understanding the processes of problem-based learning." Journal of Learning in Higher Education 13 (2) (2017).

Nietzsche, F: "The Gay Science" (1882/1887), E W Fritzsch, Leipzig.
Sartre, J-P: " Being and Nothingness", Washington Square Press, N.Y. 1956 (1943).
Schutz, A: "The phenomenology of the social world". Heinemann Educational Books, London, 1972.
Silverman, D: "The Theory of Organizations", Heinemann, 1970.
Weick, K E: "The Social Psychology of Organizing", Addison-Wesley Publishing Company 1979 (1969).

CHAPTER 2

The Individuality of Experience in Organizational Life

Abstract This chapter introduces concepts such as intentionality, situatedness, historicity, understanding, noema and noesis as it develops the epistemological and ultimately phenomenological foundation for being and organizational interactions. It introduces an understanding of the human being, and the tradition and the arguments for the world as a social and individual phenomenon. It highlights some of the important concepts of human beings used later as central for the discussion of humans and organizations, for example consciousness, intentionality, understanding, self-awareness and critique. The focus is on building an ontology and epistemology for understanding the essence of being and being together in organizations.

Keywords The subject • Consciousness • Thinking • Being • Understanding • Lifeworld • Intentionality • Historicity • Mind and body

Every understanding of an organization, or of anything, is always from a position and a perspective. This is a central and fundamental statement in that reality does not describe itself and is not a given in any way. Reality is always seen, interpreted, understood and taken for granted from a certain point of belief and tradition. The origin of this and of all experiences of reality is the human being. It cannot be otherwise.

K. M. M. Møller, M. Fast, *Investigating Being in Organizations and Leadership*, https://doi.org/10.1007/978-3-030-58138-1_2

This chapter will discuss an ontology of phenomenology and in this an understanding of the human being. It presents a discussion of a tradition and of arguments for the world as an individual and social construction. It is a world constructed in everyday life, and the "world as nothing than the world as meaning". The discussion here will highlight some of the central and important concepts of a perspective on human beings used later as central for the discussion of the being of humans and of organizational being, for example consciousness, intentionality, understanding self-awareness and critique. The focus is on building an ontology and epistemology and a dialectical understanding of the essence of being and being together in organizations. It is a discussion of cognition and understanding of humans and human being in the world.

Phenomenology is a broad tradition and covers discussions of ontology, epistemology and methodology (e.g. Edmund Husserl [1859–1938], Martin Heidegger [1889–1976], Alfred Schutz [1899–1959], Maurice Merleau-Ponty [1908–61], Jean-Paul Sartre [1905–80] and Hans-George Gadamer [1900–2002]). It is applied in different fields in the social sciences and in science of art, for example psychology, sociology and pedagogy/learning. Central to all of these fields is an interest in the subject as a human being and on what constitutes the subject as a social being.

This can be seen in Husserl (1962) and his discussion of what a human has to seek to understand the meaning of their life, which in itself is a departure from the question of being and of being as meaning seeking (see Bjurwill 1995:37). Husserl's study is a discussion of consciousness and the "natural attitude": the relationship of consciousness to the Lifeworld, the world of ordinary, everyday lived experiences (see White 1990:78). Husserl's transcendental phenomenology is based on the world that we daily live in, experience, talk about and take for granted in all our activities (see Bengtsson 1993:43–). At the same time this approach to the world is naive, because naturally we are ignorant of the possible conditions for our existence. In our daily experiences, human beings naturally and naively take for granted reality as a substance existing in itself.

The basic questions are ontological, as they raise further questions of an understanding of what the subject is, what the world is and what the subject is in relation to the world. There are no logical ontological statements saying that humans and the social world are without structure in essence (e.g. Schutz 1973b:22). On the contrary, it is the opposite. The world has meanings and structures of relevance to the subject and to the people living and acting in its context. The world is pre-interpreted in advance through commonsense constructions of meaning made in everyday life by the subject and in a social context. It is these objects of thought—the commonsense constructions of meanings and knowing—to which people connect in their actions in a context in which they feel at home, as something they think they know and act accordingly toward.

The world is at the same time both subjective and social. The world is common to us as it is an intersubjective world, as something we experience with a natural attitude as something we share or can share, and share with meanings. The world was there before us, as it existed and has its own history, and is given to us in an organized fashion in the process of socialization and learning in the context we are thrown into by birth. It is the scene and arena of our actions and the locus of resistance to action: we act not only within the world but upon the world—we act upon what we meet. It is in this commonsense world that we come into relationship with each other and try to come to terms with each other by creating meanings that we share in some situations and about different events and objects. All of this, however, is typically taken for granted, and this means that these structures of daily life are not themselves recognized or appreciated formally by common sense. Rather, common sense sees the world, acts in the world and interprets the world through these implicit typifications and interpretations connected to the subjects and their views on the situation in the context. That there is a social world, that there are fellow humans, that we can communicate meaningfully with others, that there are very broad and general principles true for daily life—all of this is interwoven in the texture of the natural attitude (see Schutz 1990:XXVII). To understand the social world is to understand the way in which human beings define different social situations and objects, and the very definition is a process and an action; interpreting the world is acting in the world. A social action is therefore an action oriented toward the past, present or future behavior of another person or persons, where the specific mode of orientating is its subjective meaning (Schutz 1972:XVII).

The Human Being and Knowing the World

As stated previously, every understanding and observation of an object is always from a position and with a perspective. Reality as such does not in itself have meaning. Meaning is connected to the thinking and acting human being. This discussion, as a statement of what is, is an ontological one and the foundation of what we are discussing when talking about being and being in organizations. Thinking is a condition for existence and being, as a condition for anything for which we can name a social context. There is no social reality without this—it is a condition that must be a priori of being, in the same way as time and space are a priori.

Consciousness and Intentionality

What characterize the human being are mind and body—thinking and acting.

The concept of **intentionality** (see Husserl 1962, Heidegger 1992, Merleau-Ponty 1994/1962, Schutz 1978a, b) is central to understanding the human being and is the foundation of an understanding of thinking, as well as being central to an understanding of the social and of organizations. Intentionality makes consciousness the defining property of the mind and orientation, but it is insufficient to define intentionality only as directedness. In experiencing an act of consciousness we find ourselves directed to something. For example, in perceiving we are directed to the thing perceived, in remembering we are directed to the event recalled, in loving or hating we are directed to the person loved or hated (cf. Gurwitsch 1982:60; Moustakas 1994:50). There is no act of thinking without an object that is thought on; no will without the willing of something; no act of judgment without something being judged… and so on.

This means that intentionality is the quality of consciousness giving meaning to the experience that is experienced. In the intentionality a subject and an object are connected: the consciousness of humans is directed to something other than itself, and this is why neither experiences nor acts and their goals can be analyzed separately. This means that when we think of something we are aware of it in a certain way. The human is the creator of the object, intentionality being the process. Intentionality is not the same as intention. It is a dimension lying behind the consciousness. What is meant here is that objects are shaped according to the way in which we observe and understand them—objects do not exist in themselves with meaning. The meaning is produced by the subject and belongs from the first moment to that subject. This is the foundation to understand the human being, and at the same time raises the question of human beings as social. This will be discussed later.

This discussion of consciousness and intentionality, this foundation, created a conception wherein consciousness and subjectivity are essential elements in the understanding of reality. One of Husserl's comments was that, in itself, (all) science is characterized by intentionality.[1] Husserl therefore encouraged philosophers to return "to the facts in themselves". In this he stressed that we can in no way acquire knowledge about objects as they are as such (in themselves). We must instead devote ourselves to objects, as they appear to **the experience**, in line with Kant's (1929) "das Ding für Uns".

Every intentionality comprises a **noema** and **noesis**, and both refer to meanings. The noema is not the real object but the phenomenon, not the thing but the appearance of the thing. The object that appears in perception varies in terms of when it is perceived, from what angle, with what background of experience and so on. From whatever angle one views an object, the synthesis of perceptions means that the thing will continue to present itself as the same real thing. The thing is out there present in time and space while the perception of the thing is in consciousness. Regardless of when or how, regardless of which components or what perception, memory, wish or judgment, the synthesis of noemata (perceived meanings) enables the experiencing person to see the thing as just this thing and no other.

Noesis constitutes the mind and the spirit, and awakens us to the meaning or sense of whatever is in perception. Noesis brings into being the consciousness of something, and refers to the act of perceiving, thinking, feeling—all of which are embedded with meanings that are concealed from consciousness. The meanings must be reorganized and drawn out. Every intentional experience is also noetic; "it is its essential nature to harbour in itself a 'meaning' of some sort, it may be many meanings" (Husserl 1962:257 in Moustakas 1994:29). In considering the noema-noesis correlate, the thing "perceived as such" is the noema; the "perfect self-evidence" is the noesis. Their relationship constitutes the intentionality of consciousness:

> *For every noema there is a noesis; for every noesis there is a noema. On the noematic side is the uncovering and explication, the unfolding and becoming distinct, the clearing of what is actually presented in consciousness. On the noetic side is an explication of the intentional processes themselves (Husserl 1977:46).*

[1] Husserl distinguished between intentionality of the act, which is that of our judgments and of those occasions when we voluntarily take up a position, and operative intentionality, that which produces the natural and anti-predicative unity of the world and of our life, being apparent in our desires, our evaluations and in the landscape we see (c.f. Merleau-Ponty 1994:xviii). It is the concept of operative intentionality that is referred to in this discussion.

> *What is meant noematically is continually changing in perception, the something meant is more, more than what is originally meant explicitly. The something meant achieves a synthesis through a continual perceiving of the whole throughout its angular visions and perceptions.* (Moustakas 1994:30)

The working out of the noema—noesis relationship, the textural (noematic) and structural (noetic) dimensions of phenomena, and the derivation of meanings are an essential function of intentionality and thereby of the being and what it is that is central to defining consciousness and being human. Moustakas (1994:31) summarizes the challenges of intentionality: explicating the sense in which our experiences are directed; discerning the features of consciousness that are essential for the individuation of objects (real or imaginary) that are before us in consciousness (noema); explicating how beliefs about such objects may be acquired, how it is that we are experiencing what we are experiencing (noesis); and integrating the noematic and noetic correlates of intentionality into meanings and essences of experience.

Merleau-Ponty's view on intentionality sticks to the original characteristics of Husserl's concept of intentionality—operative intentionality:

> *..., or that which produces the natural and antepredicative unity of the world and of our life, being apparent in our desires, our evaluations and in the landscape we see, more clearly than in objective knowledge, and furnishing the text which our knowledge tries to translate into precise language. Our relationship to the world, as it is untiringly enunciated within us, is not a thing which can be any further clarified by analysis; philosophy can only place it once more before our eyes and present it for our ratification.* (Merleau-Ponty 1994:xviii)

Intentionality is in and together with the Lifeworld. The conscious or distinct intentionality is not the original; ahead of the conscious act of thought we "intend" something. For example, I reach out my hand to something, I aim at it, not as an imagined or thought thing, but as this particular object to which I am directed: it may be a brush that I need to paint the window. The consciousness of this process and object does not have to be declared and spoken. My action is intentional as I know my bodily self and my experiences. I do not expressly think that this is a brush that must be cleaned so that I can paint the window. This deeper intentionality means that consciousness is not "I think that", but "I can". The conscious reflection builds upon a richness of preceding unexpressed intentions, and reflection is just reflection on something that precedes it.

Sight and movement are specific ways of entering into relationship with objects. And if, through all these experiences, some unique function finds its expression, it is the momentum of existence which does not cancel out the radical diversity of content, because it links its components to each other, not by placing them all under the control of an "I think", but by guiding them toward the intersensory unity of a "world". Movement is not thought about movement, and bodily space is not thought of or represented. Each voluntary movement takes place in a setting, against a background determined by the movement itself. We perform our movements in a space which is not "empty" or unrelated to them, but which, on the contrary, bears a highly determinate relation to them: movement and background are, in fact, only artificially separated states of a unique totality. In the action of the hand which is raised toward an object is contained a reference to the object, not as an object represented, but as that highly specific thing toward which we project ourselves, near which we are, in anticipation, and which we haunt. Consciousness is being-toward-the-thing through the intermediary of the body (Merleau-Ponty 1994:137–).

This can also be seen in the distinction between **"the act of thinking"** and **"the object of thought"** (Schutz 1973a:102), as it a focus on meaning and search for the underlying elements in "the stream of consciousness". This is decisive, as it introduces a temporal dimension supporting the concept of "reflexivity". **Consciousness** is an unbroken stream of lived-through experiences which have no meaning in themselves. The meaning depends on reflexivity—the process of turning to yourself and reflecting on the experience of the act—the cogito of cogito—the thinking of the thinking. Meaning is connected with actions in a retrospective way—after the action the experiences of what the act has meant for the subject can arise. This process of giving meanings reflexively depends on the actor's identification of the aim or goal that the actor tries to reach, or what could have been a meaning of the act that is now meaningful.

Understanding as Being in the Lifeworld

Understanding can be regarded as the crux of the matter, both in the ontological and the epistemological discussion. It is essential in relation to science and in the methodological discussion, and it is central in relation to humans in their creation of meaning in everyday life. In the scientific and empirical contexts, to understand is **to seek understanding of something**. A process of understanding (Gadamer 1986:109–) always remains a risk and leaves no room for a simple application of a general knowledge of rules to statements, texts to be understood, or context. Successful understanding of something means a growth in inner awareness, as a new experience enters into the texture of one's own mental experience. It is capable of contributing in a special way to the broadening of our human experiences, our self-knowledge and our horizon, for everything understanding mediates is mediated along with ourselves and our development.

Gadamer's discussion builds on the conception that our situation as human beings is that we are historical beings—we are always standing in the middle of history. The phenomenon that we wish to understand and the I who wants to understand are both related to a **context of traditions**. The aim of understanding is not only to understand the other, but also always understanding of one-self—that all understanding is ultimately self-understanding (Gadamer 1993:260). This is what we previously termed the cogito cogito—thinking of own thinking, and what this means for understanding oneself.

Understanding is something that penetrates all our experiences, because understanding is not a method, but a way of existing as a human being and orientating our self in the everyday of life. During our whole life we continue to interpret and reinterpret our experiences in life. Even memory is a continuously repeating act of interpretation. As we remember a preceding event, we reconstruct it in accordance with our present attitudes to what is important and what is not (Berger 1980:55). Understanding is just a feature of human life. Life is not something that we face, but something that we are in the middle of as life is lived life. We are in our understanding part of a historical tradition, and it is an interplay between the development in the tradition and the interpreter of that tradition.[2]

A person who is trying to understand a text always projects onto that text (Gadamer 1993:267). The subject projects a meaning for the text as a whole as soon as some initial meanings emerge from the text. The initial meanings emerge because the subject reads the text with certain expectations in relation to a certain meaning: working out this fore-projection, which is constantly revised in terms of what emerges as he penetrates into

[2] See below; cf. Gadamer (1986:100, 1993:293); Burrell and Morgan (1980:237); O'Neill (1978).

the meaning, is understanding what is there. But it is not our own meanings and fore-projections that is the focus, but rather the text. So the question will always be how to overcome and be conscious of these.

In the attempt to understand social and cultural phenomena, the observer must therefore enter into a dialogue with the subject he is studying (cf. Gadamer 1986). The understanding of a language "does not comprise a procedure of interpretation". To understand a language is to be able to "live in it"—a principle "that holds not only for living, but for dead languages". The problem is therefore not a problem of the accurate mastery of a language, but of the correct understanding of the things that are accomplished through the medium of language (Giddens 1976:55). Everything considered, when it makes sense to talk about "making ourselves acquainted with something", then we must say that we make ourselves acquainted with the writer's **meaning** of the matter that he talks about. This implies that we are trying to let the things that are said be heard with their claims to say something true.

In relation to what we have stated, as we are always situated with a position and a perspective, Gadamer talks about a "fore-understanding" (and a prejudice), which we drag into the process when we wish to understand something. To understand a text or a practice, we must through a fore-understanding form an opinion of the whole that we meet. Only on the basis of this anticipation of what can meaningfully be said to hang together is it possible for us to meet the text and interpret it in the light of the ideal of the perfect text. The fore-understanding of the interpreter is therefore part of the process of understanding, as this can be determined as the mutual conditional relationship between what the text says and the full meaning, against the background of which the interpreter assesses the text, when he construes it (cf. Gadamer 1993:269). This understanding of fore-understanding of something means that this is just a part of being as we can never escape ourselves and the way we are situated. But we can work on the understanding of ourselves and be conscious of how we understand things.

The Movement in Understanding

This discussion of the circle movement in understanding is central to Gadamer's discussion, where his objective is to discuss the universal conditions of understanding and being, through a **description of what happens in understanding** and in the being in the world. Central concepts in this discussion are prejudice, fore-conception of completeness, temporal distance, history of effect, historically effected consciousness, situation, horizon, fusions of horizons, experience and application (cf. Gadamer 1993).

Prejudice is something inherent in being human. Prejudice is the specific manifestation of the historical existence of humans as history does not belong to us, but we belong to history—the tradition. Before we understand ourselves through retrospection, we understand ourselves in a natural way in a social context. This belonging to history means that prejudice, far more than our judgments, is the foundation of our being. We meet the world as children, and socialization is simply having prejudices as the commonsense knowledge of understanding and acting in the world.

Prejudices are indispensable, as mutual understanding rests on prejudices used in interaction: (a) In a situation where we have common prejudices in seeing a problem or a phenomenon, this does not create any problems in the interaction, as we can immediately understand the other and come to terms with him. (b) In a situation where we are confronted with something new and something we do not know, we wonder about something, and we do this by virtue of our prejudices as they are different to what we confront. In this we need to enter a process of understanding, with the aim of understanding the new. Prejudices may be said to have a treble character of time: (a) they have come to us from tradition and history (before); (b) they are a constituent of what we are now and are about to be (now) and (c) they are expectant, being open to future testing and change (future). The epistemologically question is not how we avoid prejudices, in order to find a safe foundation. Rather, it is how we can distinguish between fruitful and unfruitful prejudices.

In this we must make two suppositions to begin the process of understanding. First we must have two previous expectations—**fore-conception of completeness**—the expectation of coherence and truth. An example: when we receive a text, we both expect that it is complete and understandable, and that it speaks the truth about its content. We must decide that it is possible to understand. A process of interpretation must emanate from a "pre-concept on completeness", as a formal condition of all understanding. So when we read a text we always assume its completeness, and only when this turns out to be wrong (i.e. that the text is not understandable) do we begin to suspect the text and try to discover how to attack it. Second, understanding is always understanding of a case, and in the case there are criteria for what it means to have understood. Pure and simple understanding does not exist. Understanding is always understanding of something.

The temporal distance is a productive possibility, because tradition is unbroken up until ourselves and is filled with the interpretations of the own situation of other generations. We can relate to these, and make new interpretations. In our historical situation with its special interests and problems new questions can be formulated. The context in which I am, and I with my experiences, can ask other questions and find other perspectives than those which I or others could previously. The subject itself lives in history with its new situations and requirements for action. As we ourselves must create history by forming it, we are referred to the past and to an interpretation of it. A human being who today experiences a situation as a revolutionary situation at the same time stands in a new relationship of interpretation to the revolutions about which history tells (cf. Wind 1987:65).

The above can be summed up under the concept **history of effect**: history is previous to humans and we are thrown into existence. In our understanding of history, history is already operative by virtue of the prejudices with which history has supplied us. This means that the understanding we create is always bound by the situation, and this breaks through in all processes of understanding whether we are conscious of it or not. It is to be conscious of this that will be the critical issue in understanding. We can never escape our own historicism, as we are part of history and are doomed to start a dialogue with the past, if we understand it. We can never understand the events of the past in full, as they must be understood as manifestations with a claim to truth and to being accepted or rejected as such. History is always more than an objective, present and finished past. History is also tradition and as such is binding, as it affects us. When we are part of history ourselves, this also means that we are standing in the middle of the history of the writings and stories handed down to us and we thus depend on tradition.

We are bound by the situation and this requires a development of a **historically effected consciousness**. This involves a claim in the process of understanding always to reflect on what it means to understanding/cognition/research that we are always standing in and are bound by a situation. In other words, that consciousness of something is deeply rooted in self-consciousness. In the self-consciousness, consciousness is reflexive—consciousness can withdraw from that of which it is conscious and the context to which it is immediately attached, and thus focus on itself in its difference from all other beings.

To be conscious of **the situation** is difficult, as the situation is not something we face but something we are in as situated. We cannot escape and create a distance, and can only maintain the situation as it rules in any understanding: as that which has barriers.

A situation has a **horizon** as the range of vision that includes everything that can be seen from a particular point. It is applied to the thinking mind, as we can speak of narrowness of horizon, of the possible expansion of horizon, of the opening up of new horizons and so forth (cf. Gadamer 1993:302). The horizon is thus a series of inevitable, implied concepts, theories and experiences which color our interpretation of life and the world in which we live. The horizon is in constant movement and construction through a process in which we continuously test our prejudices and reinterpret them. Horizon always appears in a context, and our consciousness never stands alone, but is in relation to others—we have a relation, and this means that we see certain connections and relations. In short, we all have some prejudices.

Understanding can thus be seen as **fusions of horizon** (ibid.:307). This does not mean leaving the own horizon and making yourself acquainted with that of another person and trying to reconstruct it, but rather taking an open and receptive attitude in order to acquire experiences, the situation considered. We draw our historicism into the understanding and in relation to the historicism of the other person. The other person talks from his horizon of meanings, prejudices and questions, and we do the same. We must continuously alternate between penetrating the horizon of the other person and linking this back to our own horizon. Understanding has this dialectic character in the interaction between the person who interprets and the meaning formed. In other words, through this fusion of horizons the other person and I will reach a common horizon—and at the same time I apprehend the other person in the person's own peculiarity. Here we therefore cannot talk about a correct interpretation or meaning; it is about openness and change in understanding, as everyone has an own horizon and perspective in their understanding. The process of experience may thus be understood as change; that we change (through self-cognition, development of the consciousness, etc.), that the phenomenon acquires another interpretation and meaning.

The Lifeworld: The "I" Being in the World

The Lifeworld is understood as the immediately experienced world, as it appears before it is subjected to a scientific investigation. It is the historical reality from which the human being immediately takes their bearings. This is the reality we live in every day as we understand the organization in a certain way—as taken for granted in its meanings and patterns. This reality can be understood as W. James stated: reality simply means relation to our emotional and active life. The origin of all reality is subjective—all that titillates and stimulates our interest is real. To call a thing real means that this thing stands in a certain relation to us. The word "reality" is in short a frame (Schutz 1973b:60). In other words, reality is linked to the subject, as it—that something—appears to experience it, and from this we have the process of creating meaning, of that which appears to the subject.

Gadamer considers the concept of **Lifeworld** as the antithesis of all objectivism. It is an essential historical concept that does not refer to a universe of being, to an "existing world". Nor can the infinite idea of a true world be meaningful when created out of the infinite progress of a human historical world in historical experience. It is not this conception of the world that natural science tries to imagine or to acquire knowledge of. Lifeworld means something else, namely the whole in which we live as historical creatures in the everyday. And here we cannot avoid the consequence that, given the historicity of experience implied in it, the idea of a universe of possible historical life worlds simply does not make sense. It is clear that the Lifeworld is always at the same time a communal world that involves being with other people as well. It is a world of persons, and in the natural attitude the validity of this personal world is always assumed (Gadamer 1993:247).

The understanding of Lifeworld is central in the perspective. It is in this Lifeworld that one must relate and understand all human activities, and on which scientific and philosophic understanding have to be based and reflected upon and from where it has its departure. However, the argumentation in this is different in the various philosophical traditions.

Merleau-Ponty argued that the world is before all analyses. I am not a "living creature", nor even a "man", nor again even "a consciousness". I am the absolute source of the world (as given meaning to it). My existence does not stem from my antecedents, from my physical and social environment. My existence moves out toward them and sustains them. For I alone bring into being for myself (and therefore into being in the only sense that the word can have for me) tradition which I elect to carry on. Or the horizon whose distance from me would be abolished—since that distance is not one of its properties—if I were not there to scan it with my gaze (Merleau-Ponty 1994:VIII–).

The Lifeworld is the world that is livingly present in our experiences and which is therefore indissolubly bound to the experiencing subject of the human in the world. It is in this world that we know ourselves, as it is the only world the subject knows of. The world is not what I think, but what I live through. I am open to the world, I have no doubt that I am in communication with it, but I do not possess it; it is inexhaustible. "There is a world", or rather "There is the world"; I can never completely account for this ever-reiterated assertion in my life. This facticity of the world is what constitutes the *Weltlichkeit der Welt*, what causes the world to be the world; just as the facticity of the cogito is not an imperfection in itself, but rather what assures me of my existence (Merleau-Ponty 1994:XVI–).

Inversely, it is also true that the world is indissolubly bound to a subject. The only world we know of is the world that is available to us as experiencing subjects. The Lifeworld consequently turns out to be both pre-objective and pre-subjective. It is exclusively from abstractions from the Lifeworld that we can talk of a pure nature and a pure subject, respectively. With that, Merleau-Ponty means that the Lifeworld represents a third dimension as agent between naturalism (objectivism) and subjectivism by building a bridge between contrasts. The reason is that on one side, the Lifeworld is a world that transcends the subject, but which at the same time is an experienced world, that is, a world connected to a subject. One can say that it is a circular relation between the world and the subject: the subject is marked by the world and the subject marks the world.[3]

Heidegger's conception of the Lifeworld appears as he abolishes the "I" and introduces *Dasein* (Being) in the understanding of the subject and

[3] This conception of Merleau-Ponty originates in his early inspiration from dialectics (from a Marxist conception) and a configuration of contrasts: for example, individual—society, nature—culture. This dialectic conception is also seen in the work of Berger and Luckmann (1966) and Silverman (1983), among others. A dialectic conception is generally incorporated in hermeneutics in relation to interpretation and understanding, for example understanding in Dilthey's hermeneutic circle—the whole-part movement in the process of interpretation. We can also find it in phenomenology as here with Merleau-Ponty, and in a discussion of dialectic of the subjectivity of the experience and the continuity of the object (e.g. Gadamer 1993:224). There are, however, different conceptions of dialectics in the different traditions.

of existence: "Dasein always understands itself of its existence – in terms of a possibility of itself: to be itself or not itself" (Heidegger 1992:33). Any being which is similar to humans relates to its own being (existence), as it is aimed (intended) at the surrounding being in the background of an understanding of the world, in which this being exists, and this being in the background of which the being can be interpreted. In other words, *Dasein* as a human form of existence is different from anything else that exists in the world. This means that *Dasein* is the subject as worldly, as being, as existence, as situated in the world and as agent—as acting in the world. "World", in this existential sense, would be found in the self-reflective consciousness even of a rather primitive awareness, for which the limits of the world may well be the limits of a village or country. It is, in this sense, the most general concept about existence: the place in which one is. Dasein, being in the world, Heidegger simply understands as being **in** the world, as a feeling, through experience, of the world as familiar. In other words, the world is something that we know and feel safe about and which constitutes our "home". To-be-in-the-world is the ultimate presupposition of knowledge (this places ontology prior to epistemology). The bases of epistemology are the knower and the known. But prior to the distinction between knower and known (or subject and object) is the fact that the subject can relate to a known, which means that the presupposition of the very subject—object distinction is grounded in an already admitted basis of relationship. That is, that the subject has a world in which the object can occur. Knowledge does not occur in isolation from one's world of concern and environment (Gelven 1989:60), as it can only be produced in this world as the subject is situated in its being.

The way of orientating in the world of the subject and finding meanings is discussed by Schutz in his reflections on commonsense knowledge of the world. He stressed that the world is not a world but worlds, as he discusses the understanding of the world around us as varying from context to context. In other words, we live in a world full of multiple realities and where each of them is defined through "finite provinces of meaning" (i.e. context-dependent meanings). Schutz talks about provinces of meaning because it is the meaning of our experiences and not the ontological structure of the objects that constitutes the context. This is the discussion of intentionality, as the subject directedness and thinking of it is the source of the experience. Each province has its own cognitive style with respect to which experiences within each world are inter-consistent. And each of the finite provinces of meaning may receive the "accent of reality", and may be attended to as real. Not only are the images of reality of each individual different, but also when the actor changes the social context, there are other basic rules for the individual, dependent on whether he is at work or at home. These changes ("leap of consciousness") make it possible for actors to overcome differences between different worlds—it is the question of being aware of the experience and the world, but also what was discussed earlier as the cogito cogito—thinking of own thinking.

The core social relation is directed toward the "We-relationship", and all other notions of social forms that are applied by actors in their everyday social life are derived from this. In any face-to-face encounter, the actor brings to the relationship a stock of "knowledge in hand", or "common-sense understandings". Thus the actor typifies the other actors; he is able to calculate the probable response of the other to his actions, and sustains communication with him. An actor's "stock of knowledge" is taken for granted as "adequate until further notice". Its totality is composed of "self-evidences" changing from situation to situation and being set in relief at any given time by a background of indeterminacy (cf. Giddens 1976:29). If we consider the everyday working Lifeworld, it is a "finite province of meaning" among many others, though it is emphasized as the extreme and all-important reality.

In respect to the reality of everyday life, we are, in the natural attitude, made to take part in it because our practical experiences prove that the unity and congruity of the labor (work) worlds are valid and that the hypothesis on its reality is irrefutable. This reality seems natural to us. Furthermore, we are not ready to leave our natural attitude toward it without having experienced a certain shock that forces us to break through the boundaries of this finite province of meaning and change the accent of reality to another. We have the same world of directly experienced social reality in common: the world that surrounds my Here and Now corresponds with that which surrounds the Here and Now of other people. My Here and Now includes that of the other person, together with his attention to my world, in the same way as the content of me and my consciousness belongs to the world of the other in his Here and Now. However, this domain of directly experienced social realities is only one of many social fields. In the same way as my actual perception is only a fragment of the world of all my experiences, and hence this is only a fragment of the entire world of possible experiences. Seen in this way, I can directly experience the social world in fragments, as I live from moment to moment (Schutz 1972:142).

This discussion and reflection on phenomenology and of the existential in being human leads to a discussion in the next chapter of how we can understand the subject as a being and the becoming of a social being in the world.

References

Berger, P L.: "Invitation til Sociologi – ett humanistisk perspektiv". Rabén & Sjögren, Stockholm, 1980.

Bjurwill, C.: "Fenomenologi". Studenlitteratur, Lund, 1995.

Bengtsson, J.: "Sammanflätningar – Husserls och Merleau-Pontys fenomenologi". Daidalos, Uddevalla, 1993.

Berger, P L & Luckmann, T: "The Social Construction of Reality – a treatise in the sociology of knowledge". Doubleday & Company, New York, 1966.

Burrell, G & Morgan, G: "Sociological paradigms and organisational analysis". Heinemann, London, 1980.

Gadamer, H-G: "Reason in the Age of Science". The MIT Press, Cambridge, Massachusetts, London, England, 1986.

Gadamer, H-G: "Truth and Method". Sheed & Ward, London, 1993 (1960).

Gelven, M.: "A commentary on Heidegger's being and time". Northern Illinois University Press, DeKalb, Illinois, 1989.

Giddens, A.: "Hermeneutics, Ethnomethodology, and problems of interpretative analysis". In L A Coser & O N Larsen (eds): "The uses of controversy in sociology", The Free Press, New York, 1976.

Gurwitsch, A.: "Husserl's Theory of the Intentionality of Consciousness". In Dreyfus, H L. (ed): "Husserl Intentionality and Cognitive Science", The MIT Press, Cambridge, Massachusetts, London, 1982.

Heidegger, M.: "Being and time". (1927) Blackwell, Oxford, 1992.

Husserl, E.: "Ideas". Macmillan, New York, 1962.

Kant, I. "Critique of pure reason" (*"Kritiken der reinen Vernunft" (1781/1787)*). Macmillan, Hong Kong, 1929.

Merleau-Ponty, M.: "The Phenomenology of Perception". Routledge & Kegan Paul, London, 1994 (1962).

Moustakas, C: "Phenomenological Research Methods. Sage, California 1994.

O'Neill, J.: "Can phenomenology be critical?". In Luckmann, T (ed.): "Phenomenology and sociology", Penguin, London, 1978.

Sartre, J-P: Being and Nothingness", Washington Square Press, N.Y., 1956.

Schutz, A: "Hverdagslivets sociologi". Hans Reitzel, København, 1973b.

Schutz, A: "Collected Papers I: The problem of social reality". Kluwer Academic Publishers, The Netherlands, 1990.

Schutz, A: "The phenomenology of the social world". Heinemann Educational Books, London, 1972.
Schutz, A: "Concepts and Theory Formation in the Social Sciences". In Emmet D & Macintyre A (ed.): "Sociological theory and philosophical analysis". The Macmillan Company, New York, 1978a.
Schutz, A: "The theory of social action". Indiana University Press, Bloomington and London, 1978b.
Schutz, A: "Some leading concepts of phenomenology". In Collected Papers I: "The problem of social reality". Matinus Nijhoff, Haag, 1973a.
Silverman, D: "The Theory of Organisations". Heinemann, London, 1983.
White, J D.: "Phenomenology and Organizational development". Administrative Science Quarterly, Spring 1990.
Wind, H C.: "Historie og forståelse—filosofisk hermeneutik". Aarhus Universitetsforlag, Aarhus, Denmark, 1987.

CHAPTER 3

The I and the Becoming of Social Being

Abstract This chapter analyzes to what extent the human being can be considered a social phenomenon, an intersubjective being. It discusses different perspectives and details regarding intersubjectivity, reflexivity and empathy in order to establish an understanding of the ontological functions of human beings.

Keywords Consciousness • Intersubjectivity • Storytelling • Apperception • Thinking

To describe humans as a general concept is an ambitious aspiration, for many reasons. One of the primary challenges with this aspiration is that initially it assumes that the writer can somehow detach him/herself from the unique experiences and emotions of being a specific human being. In contrast to a biologist observing frogs, we humans have our own experiences and perspective of what it means to be human. These experiences are all we have, which means we cannot look elsewhere for more valid sources. However, as shown by Kant in his concept of transcendence, Descartes in his method of doubt and Eco in his analysis of the limits of language, perhaps there is something in our own experiences which can provide validity about the lives of others and therefore say something about humans as a general concept (Eco 2000; Kant 1929; Descartes 2007). All three methods aspire to analyze the individual experience in order to describe a

K. M. M. Møller, M. Fast, *Investigating Being in Organizations and Leadership*, https://doi.org/10.1007/978-3-030-58138-1_3

phenomenon, which pertains to more than the individual experience. All three analyses face the same challenge of providing a method for differentiating in our experience between that which is primarily originating from ourselves and that which primarily originates from the experienced phenomena, as well as showing which parts of which originating elements can be valid for more than the origin. Different elements can theoretically be assumed: the a priori elements of experiencing something, the process of experiencing something, the result of experiencing something. These three distinctions rely on the assumptions that everything has something prior to it that enables the existence of it; that everything has a time and place; and lastly that everything humans can experience exists. An example of all this is the notion of a thought. For adults at least, thoughts occur in language and they occur as a finished product, the manufacturing process of which we cannot experience. I cannot know what I am thinking until I have made that thought. In that sense, our thoughts are not intended by our consciousness, but rather they happen *to* our consciousness. One might say that our consciousness can direct itself toward our thoughts. This means that we can experience our thoughts only as a result of a process and that process must have a priori elements which enable it, for example the ability to create thoughts. In other words our thoughts have all the required elements to exist for us, but not all of those elements are available to us. This is the case for a vast number of things in our lives. Our perception captures the result of things coming into our experience, but not the process. Moreover, this notion of things happening to us, which is commonly defined as something we do, an action we perform, will be highly relevant later. For now the question remains: What is a human being and how can such a thing be described with any sense of validity? What about the sentences I am writing now? Are they happening to me? Or are they performed and planned by me? It all depends on the definition of "me". The beginning to an answer is obviously that "me" or humans have to be defined as something which can both be happening and performing. We can in other words carry out actions which are not consciously intended but rather unconsciously occurring, and we are obviously able to carry out actions which have been intended prior to the execution of the actions. Furthermore, it is necessary to include both consciousness and unconsciousness in the definition of me. In other words, I am more than I think.

> *But only by taking the attitude of the generalized other toward himself, in one or another of these ways, can he think at all; for only thus can thinking – or the internalized conversation of gestures which constitutes thinking – occur.* (Mead 1967:156)

In the text above, Mead indicates a strong dialectical relationship between the inside and outside of a person due to the construction of thinking as a human ability. Thinking is usually considered highly internal but nonetheless highly affected by the social and external context. One can argue that Mead exemplifies the problematic elements of the dichotomy of internal and external through the usage of that very same dichotomy. For a moment, let us suspend the common belief that you are you, and only you. Just as you are thinking and only you are performing your thinking. What would happen if we instead consider thinking and ourselves as the result of a vast amount of complex experiences, which result in a particular way of thinking in a particular person, such as you? Imagine two puddles of water next to each other on a rainy day. Both puddles increase in size until ultimately one puddle overflows into the other and the puddles intersect. How do you determine where one puddle ends and the other begins? Are they just one puddle now? Or have there never been two but instead only one puddle which had not yet connected? The same challenge is one posed for the description of humans. If we accept this thinking, as will be shown, all of human experience and characteristics originate not solely from ourselves but rather primarily from our social co-existence or the many other puddles. Where does one human begin and the other end? The point of this chapter is to show how existentially interconnected and dependent we are and how such an understanding can critique and develop some of the dominating organizational and economic assumptions.

Let us return to the process of thinking as it pertains to consciousness and view it as the act of experiencing thoughts happening. As Mead, Pierce, many others and our own experiences point out, the process of thinking happens in the form of the semiotic system of language (Pierce 1992, 1998). Briefly, language can be described as a system of symbols, which contains a vast number of references to culturally derived experiences. The language, like any medium, only enables the type of information which is possible through the usage of that specific medium. As such, language and thoughts can serve as a tool for some forms of human communication. And how can we describe that which is not available through the usage of language? The details and nuances of our emotions, for example? The encounter with new and confusing emotional experiences of which our descriptions are merely crude approximations. We argue that the existence of a category of experience which cannot be described through our current medium of language is as certain as it is unattainable through that very same language. As such, any single example or attempt at describing the category will fall short, and all we can do is point the reader's attention to any such experience that that reader might recall (Møller 2017b). It might possibly be that philosophers in their attempt to understand the human experience should look toward the artist and poets for an understanding of whatever does not lend itself to language, since the work of artists and poets is intended to create emotional responses rather than the scientific aspiration to create almost solely conscious rationality. We are aware that our discussions regarding intentionality contradict the idea of rationality without emotions, which indeed is a significant point in this book. It is in fact the existence of ambitions of emotionless rationality which justifies our very critique.

In much the same way as the category of experience, language in itself is therefore understood as a way of pointing toward some experience with the hope that the experience we point toward is somewhat shared by the person we are directing. This is abundantly clear when it comes to interactions and the alignment of assumptions necessary for establishing any kind of relationship. The aspects of intentionality described in the previous chapter are central to any understanding of interaction and development, and our understanding of the other even plays a central role in our understanding of our contexts. Husserl establishes our understanding of the actions of the other as an essential part of the continuity of our perception (Husserl 1995). Our very notion of objectivity and its many applications, such as stability and physical laws, rests on our ability to epistemologically incorporate the other in our way of functioning. Weick describes one of the consequences of this in relation to our understanding of identity:

> *(...) who we think we are (identity) as organizational actors shapes what we enact and how we interpret, which affects what outsiders think we are (image) and how they treat us, which stabilizes or destabilizes our identity. Who we are lies importantly in the hands of others (...)* (Weick and Obstfeld 2005:416)

The quotation from Weick is here interpreted as the "hand of others" being the accumulated experience throughout life, not meant as single individuals always possessing the power to reshape our identity (nor do we think Weick intended it so). The reason why the essentiality of the other is relevant for our discussion of language is that both language and the other play a vital role in our ability to understand a phenomenon. As described earlier, phenomena can refer to everything that is available to our experience and this entails only our attempt to understand the essence of a certain phenomenon. Language is our way of communicating with ourselves and others and therefore it serves as the primary means of improving our ability to analyze a phenomenon. If language is our way of pointing at a phenomenon, the other is the intention of our pointing; for individuals, the other constitutes the essence of ourselves and the phenomena we are pointing toward. So far, we have discussed how the other affects the individual, but the dialectical process of interaction enables the individual to likewise affect the other; and in order to understand the becoming of a social and organizational being, the result of such a dialectical process is highly relevant.

> *The individual is no thrall of society. He constitutes society as genuinely as society constitutes the individual. (…) Human society as we know it could not exist without minds and selves, since all its most characteristic features presuppose the possession of minds and selves by its individual members; but it's members would not possess minds and selves if these had not arisen within or emerged out of the human social process (…)* (Mead in Clark & Fast 2019:143)

Based on Mead's description, the way in which we understand the other is essential to understanding not only society but also ourselves. The process of understanding rests entirely upon one's ability to empathize with the other. There is an interesting example in neuroscience regarding mirror neurons and their role in observing others, which hopefully will prove quite intuitive and relevant for grasping the philosophical notion of empathy.

> *All these studies suggested that humans have a 'mirror matching system' similar to that originally discovered in monkeys. Whenever we are looking at someone performing an action, beside the activation of various visual areas, there is a concurrent activation of the motor circuits that are recruited when we ourselves perform that action. Although we do not overtly reproduce the observed action, nevertheless our motor system becomes active* ***as if*** *we were executing that very same action that we are observing. To spell it out in different words, action observation implies* ***action simulation***. (Gallese 2013:37)

In this quotation Gallese makes two conclusions which are philosophically significant. First, observing someone involves ourselves far beyond merely our visual engagement. It involves selves and our experience of doing something similar to what we observe. In other words, our empathetic understanding of the other is created and limited by our experiences of similar events performed by us. Second, empathy or observation as noted here is an action, and therefore its precision and efficiency depends on our practice of said action. As with any action, an understanding developed through empathy is therefore also the result of our desire to perform the empathetic action. From a philosophical origin Crossley described in his discussion of Husserl the two concepts Pairing and Apperception as a detailed description of the process of observation to understanding (Crossley 1996:6). Here apperception is the notion of completing the mental image from what has been observed—when we walk by a house we assume the existence of the sides of the house which we cannot see because we have seen numerous houses which fit our concept of a house having four sides. Pairing is the process of establishing a link between the observed phenomenon and an appropriate mental concept—we determine an object as being a house. From there we use apperception to infer the non-visible sides of the house. The understandings of empathy from Gallese, Crossley and Husserl are similar with regard to empathy as action which resets entirely our ability to infer characteristics of a perception based on previous experience.

Regarding empathy as an action we perform, it is important to note that, unlike other actions, we cannot choose not to perform empathy. We do so every time we walk past someone on the street or every time we see someone in the break room. The extent and result of empathy can vary greatly but it is a constant in interactions with other people.

Implicit in the discussion of thoughts, empathy and perception is the notion of individuals unconsciously and inevitably selecting relevant experiences when interpreting a phenomenon. Experiences as a concept describe prior events and our responses to those events as they are recollected in relation to a phenomenon. Experience therefore consists of more than merely a descriptive account of a timeline of events. They consist of our emotional response, which in turn constitutes the framework of our responses toward the phenomenon which instigated the recollection of those experiences and emotions. This response can obviously occur consciously but the more relevant and frequent process is the unconsciously interpretational process, which takes places in the same constant fashion as empathy. In regard to the notion of constant performance, the act of emotional perception is no different than the act of empathy. In fact, because we cannot observe consciousness, and can only infer it from observation, the notion of interpreting a house as consisting of four walls appears no different from assuming the consciousness of another. Both processes are limited by our own experiences and ability to process those experiences imaginatively. Understanding our entangled state of existence with the other, from puddle to puddle, becomes a question of understanding and developing our current limits given by our experiences or lack thereof. We can never become the other puddle, but we can become a far bigger puddle if the desire to understand the other and critically develop ourselves exists. In order to do so it is important to recognize what we can choose what happens to us and our cognitive and emotional limitations.

From this perspective, humans or organizational members can be misinterpreted to be solely the product of social interactions, of cultural development or, as some state, tabula rasa. Given the argumentation from Chap. 2 and the beginning of this chapter, it is clear that the individual is to a large extent the product of social interaction—we learn from our current contexts. However, as with the water in the puddles, individuals have a set of functions which are initially without content but in principle constant over time. An example using the water analogy is its ability (or function for humans) to always move toward the lowest point it can find regardless of where it is, what it hits and so on. Likewise, we can always observe, recall and feel but the content of our function is what varies depending on where we are and what we encounter. These functions are not binary, as in I can recall or I cannot recall. We can train our ability to recall or we can let it decay, and it is because of that fact that tabula rasa is not an adequate description of human beings. The choice to recall and the content of a recollection is a product of our interpretation of our social context, and therefore the functions and their ability to efficiently process content are also a product of the content of those functions. In other words, the blank slate is not really a slate if the edges, color and other characteristics change depending on what you write on it. It is not a matter of nature or nurture as the separation of these two elements is merely a theoretical division, which does not adequately describe reality. The same critique goes for the discussion of rationality vs emotions if one considers rationality an aspiration of logical argumentation as an opposition to emotional behavior. As stated earlier, no behavior or thought can occur without emotions, even those we consider rational. A more meaningful description could be that we are emotional beings who in order to convincingly communicate in a structured manner have created a cultural agreement regarding a specific set of values for our communication. These values are still within the perceptual limits of ourselves, and the methodology they idealize has proven adequate over time. The same can be stated in regard to the concept of professionalism as an aspiration to act in the same manner regardless of the individual present in the interaction. Professionalism refers to a set of values which are much more contextual and less proven over time, within a certain context, and demands that all act in the same accordance to those values. There can be many practical challenges to such a set of values which depend on the context, and therefore the potential challenges are far too numerous to address here. However, as stated earlier, there can also be existential challenges in the form of difference between the aspired values in, for example, professionalism and the functions of individuals. Once again, emotions and their unavoidable role in our existence is disregarded. A counter-argument to this understanding could be that the very idea of civilization includes an aspiration to disregard emotions, for example we are not allowed to kill even when we are extremely mad. But we are allowed to be mad, and we are allowed to show it. The idea in professionalism is not merely to take the right course of action, but rather each organizational member is aspiring to *be* professional. Herein lies the difference. The function of emotions is present and accounted for in civilization, whereas in modern professionalism the function is considered something that can and should be disregarded. Aspiring to be professional as it is defined here is to some degree the same as aspiring to be a machine—it is impossible, and in this manner the very aspiration will certainly be problematic.

The Intersubjective Puddle

Despite the constant nature of our functions, it is not reasonable to indicate that they can be isolated from the effects of their content, and therefore this dialectical relationship is henceforth referred to as the essence of being a human being. The following question can thus be asked: In what way do our functions then allow our values to be shaped? How do we evaluate when a value contradicts our essence?

These questions touch upon the foundation for our organizational behavior. How organizational members shape values and become critical of values presented to them are essential aspects of the formation, development and ultimately efficiency of organizations. The alignment of actions is entirely dependent on a sufficiently similar interpretation of a given problem. How we address and solve a problem is entirely dependent on how we identify and evaluate that very same problem (Møller 2017a). In organizations, as in the rest of the Lifeworld, interpretation is not an action shaped only by the performer and their experiences, but also by the phenomena and their relationship with the context. Therefore, the notion of intersubjectivity becomes highly relevant when addressing organizational problems and cultural development in general. We are not solitary puddles but rather highly complex social entities with endless ties to a never-ending interpretation of our context. This is where intersubjectivity can guide the understanding of values and the essence of being a human.

Intersubjectivity has been discussed numerous times and evidently there are several issues with the concept which most prominently serve as a guiding tool away from solipsism toward a communality of truth and understanding. As Reich (2010) points out, intersubjectivity is intended to provide an answer to the question of how we can understand each other given that our ability to understand is created entirely from our own unique experience and that the consciousness of others cannot be observed, only inferred. We argue that this question makes some assumptions that will inevitably force anyone who attempts to answer it into a set of contradictions. One of these assumptions is that communication succeeds as a result of understanding –particularly because we cannot observe consciousness or intentionality, we can never claim with absolute certainty to understand the other. Understanding and interpretation are about approximation. The second assumption is about the constitution of our own ability to understand, and therein lies the solipsistic notion that our experience is ours and ours alone. As already discussed in this chapter with our puddle analogy, human beings are not solitary creatures; rather, and quite contrarily, it can be difficult if not impossible to draw the philosophical line between what in me enables the experience and what coming from other people enables the very same experience for me. Our conscious thoughts, as in the self-dialogue we can have sometimes, is not the singular cause of our everyday experience, nor does it constitute our whole being, but rather it is a part of ourselves, and if only based on frequency, it seems to be a rather small part. Therefore our self-understanding and our conscious narratives are not the only part of our ability to understand the other. The solipsistic notion which could be interpreted as a part of the intentionality concept falls short of grasping our co-existence and intertwined part in each other. Human beings can understand each other because our experiences are constituted by one another in a constant dialectical and existential relationship with other beings (whether present in context or mind). Intersubjectivity is therefore discussion of our entangled state of being and its consequences for understanding and truth. We argue that the very notion of ourselves and the other, of me and you, is a theoretical division that implies a conceptual division in experience which does not adequately emphasize the role of interaction in the creation of our existential functions. We might start out being ourselves, but we are entirely created by others. We therefore view the individual not as a solitary "I", but rather as an endless number of social processes which are made possible only by the existence of other individuals who are likewise constituted by those very same processes. Intersubjectivity is closely linked with intentionality and, in fact, one could argue that, philosophically speaking, it only serves to emphasize that which is already a part of our intentionality. Given the vastness of the concept of intentionality, it is challenging to disagree, although there might be a contextual and philosophical argument to maintaining it given a specific existential function. As discussed earlier, empathy and its ties to understanding the other play a central role in our interaction with other people. In the concepts put forth by Buber regarding the two existential relationships with others one can enter into, I-Thou and I-It, there is a function which has empirical and organizational relevance and enables an understanding of one of the essential characteristics of different organizational communities:

> *If I face a human being as my thou, and say the primary word I-Thou to him, he is not a thing among things, and does not consist of things. Thus human being is not He or She, bounded from every other He and She, a specific point in space and time within the net of the world; nor is he a nature able to be experienced and described, a loose bundle of named qualities. But with no neighbor, and whole in himself, he is Thou and fills the heavens. This does not mean that nothing exists except himself. But all else lives in his light.* (Buber 1958:15)

The I-Thou relationship describes a context wherein individuals are truly intertwined to the point of no experience or thoughts. They are there, together, for a moment in an immediate and self-forgetting fashion. One might be tempted to argue that they become each other, almost like puddles flowing into each other—one might also stretch this philosophical position too far. It is, however, a relationship which exists in significant contrast to the other intersubjective relationship, I-It. The I-It relation describes a context wherein an individual considers another individual as an object with clear limitations intended for experience or use. In regard to experience Buber further states:

> *I do not experience the man to whom I say Thou. But I take my stand in relation to him, in sanctity of the primary word. Only when I step out of it do I experience him once more. In the act of experience Thou is far away.* (Buber 1958:15)

In regard to Buber's relations three assumptions arise. The first is that both relationships are exhaustive for our potential intersubjective interactions. We can do one or the other, sometimes in very quick succession. The second is that intersubjectivity is also an attitude toward the other which is created by our empathetic and immersive abilities. The third assumption is discussed by Buber below:

> *But you believe then in the existence of a paradise in the earliest days of mankind?*
>
> *Even if it was a hell – and certainly that time to which I can go back in historical thought was full of fury and anguish and torment and cruelty – at any rate it was not unreal. The relational experiences of man in earliest days were certainly not tame and pleasant. But rather force exercised on being that is really lived than shadowy solicitude for faceless numbers!* (Buber 1958:26)

The third assumption is that the two existential relationships are expressions of the essence of being a human being. As Buber states, in most situations it is harmful to the other and oneself to remain in the I-It relationship. Buber implies that the I-it relationship is overly encouraged in modern societies as it diminishes one's ability to be immersive. Once again, the discussion of emotions and professionalism is relevant as an example of the essence of being a human versus the ideals of certain parts of organizational life, for surely one cannot be immersive and forgetting of oneself if one is to be a professional being.

Coming back to the question of organizational values, their appropriately critical evaluation is necessary to transition a bit. So far we have discussed the intersubjective and intentional nature of human beings and emphasized the importance of the I-Thou relationship. Moving forward, it is necessary to understand the dialectical nature of the I-It and the

I-Thou relationships and their importance in creating a precise description of the Lifeworld and the interactions therein. Without entering into a long account of Bourdieu, one might simply state that existence is co-existence and therefore relative to something else (Bourdieu 1989). We position ourselves and find meaning based on our relation and relative position to someone else. I am tall because I am taller than other people. I am me because I am not you. I am kind because I am kinder than most. I am considerate because I act in accordance with a perceived set of values called consideration. Over time I embody a certain set of qualities because I interpret myself in relation to my contexts and deem that these qualities can identify me. As existence is dialectical so is identity and values, and therefore both are developed in accordance with a context which can challenge and test us. As Weick stated: "Who we are lies importantly in the hands of others (…)" (Weick and Obstfeld 2005:416). This discussion is empirically exemplified in the quotation below:

> *Habitus thus implies a "sense of one's place" but also a "sense of the place of others." For example, we say of a piece of clothing, a piece of furniture, or a book: "that looks pretty bourgeois" or "that's intellectual." What are the social conditions of possibility of such a judgment? First, it presupposes that taste (or habitus) as a system of schemes of classification, is objectively referred, via the social conditionings that produced it, to a social condition: agents classify themselves, expose themselves to classification, by choosing in conformity with their taste, different attributes (clothes, types of food, drinks, sports, friends) that go well together and that go well with them or, more exactly, that suit their position.* (Bourdieu 1989:19)

The notion of schemes of classifications as objectively referred to is an interesting exemplification of embodied or known values. As with identity, values are not merely ideals which are sought after; they are also relational and therefore in a position which distinguished itself by the relation to another position. This relational notion of values, identity and relative position is in Bourdieu's work a discussion of Symbolic Capital, Field and Doxa (Bourdieu 1989). In this chapter it is merely relevant to conclude that humans are striving toward certain values and that specific values offer a specific position in a given context, which in turn serve as a part of that person's identity. In regard to this description, identity becomes the result of one's ability to navigate social contexts, along with one's ability and desire to direct one's attention toward oneself. As a closing remark regarding the intersubjective essence of being human, a function which has not featured as prominently in this chapter as its importance merits is language. The remainder of this book could be a discussion of language and still not be fully comprehensive. As this chapter is rather a discussion of the social existence in which we find ourselves, language as briefly defined earlier in this chapter can be understood as the semiotic system we use to express ourselves and interpret our context.[1] As mentioned earlier in the quotation by Mead (1967), thinking is an internal conversation of gestures. In this regard gestures and symbols are considered synonymous. In other words, our internal process of thought, our self-dialogue, is performed within the same semiotic conditions as our gesticulations with other people. Therefore, thinking can be viewed as self-dialogue, and as such, thinking as well as conversations with other people are stories. The storytelling literature is vast and features various differing positions and understandings.[2] In this chapter and for these discussions stories are merely gesticulatory processes, which contain assumptions and values. The concept of storytelling is in common usage describing a process which can be chosen or disregarded. We can choose to tell a story or we can choose not to. In this version of storytelling, it is something that we do constantly, to ourselves and to others. Every statement we can make includes assumptions and values and every statement is a process we perform. Therefore, any gesticulation we might make is an act of storytelling. It might not be the entire story as we see it ourselves, but it is a story or part of a story either for ourselves or for someone else. Stories are therefore a function of our ability to express ourselves and in turn function as a guiding tool for our perception of ourselves and others. In regard to the previous discussion about intersubjectivity the quotation below is an example of how we interact and understand others.

> *There are times, however, when stories crystallize around particular interpretations. Different versions may diverge in numerous details but seem to agree on the story's core symbolism. This symbolism seems very powerful, the stories being treated as part of the heritage of an organization or of a group.* (Gabriel 2000:42)

The stable content of ourselves and our understanding of reality is therefore not given by some objective fact in itself, but rather it is a

[1] Here we rely largely on the work of Charles Sanders Pierce.

[2] For interesting works, see David Boje (2008) and Kenneth Mølbjerg Jørgensen (2017).

product of our experience, functions and position relative to the phenomenon we encounter in a given context. The ability to tell stories of ourselves to others and interpret their reaction is therefore central for understanding in general and also for identifying ourselves. Understanding is in essence the process of gesticulating the result of one's co-existence with a particular phenomenon as it appears given the current experience and contextual position of oneself. Central to this gesticulation and development of both understanding and identity is that all three elements are based upon the concept of dialectics. Therefore this concept and how it pertains to the understanding of organizations is essential to any analysis of organizations and organizational life.

References

Boje D. M. 2008: "Storytelling organizations". London: Sage Publications Ltd.

Bourdieu P. 1989: "Social Space and Symbolic Power". Sociological Theory. Vol 7. nr. 1. Pp.14–25.

Buber, M. 1958: "I and Thou". Continuum, London. 2. Edition. Translated by Ronald Gregor Smith.

Clark, W Woodrow II and Fast, Michael: "Qualitative Economics – The science of Economics", Springer, Cham, Switzerland, 2019.

Crossley N. 1996: "Intersubjectivity. The Fabric of Social Becoming". Sage Publications London, 1. Edition.

Descartes R. 2007: "Discourse of the Method of Rightly Conducting One's Reason and seeking truth in the Sciences". By Jonathan Bennett. http://www.earlymoderntexts.com/assets/pdfs/descartes1637.pdf.

Eco, U. 2000: "KANT and the platypus – Essays on Language and Cognition". Vintage. Milan.

Gabriel, Y. 2000: "Storytelling in Organizations – Facts, Fictions, and Fantasies". Oxford University Press.

Gallese V. 2013: "The 'Shared Manifold' hypothesis – From mirror neurons to empathy". Imprint Academic. Journal of Consciousness Studies.

Husserl, E. 1995: "Cartesian Meditations". Kluwer Academic Publishers. Holland.

Jørgensen K. M. 2017: "Vibrant power, vibrant subjectivities: A storytelling approach to the study of power in education". Universidade do Vale do Rio dos Sinos. Vol 21. Issue 1. Pp.21–30.

Kant, I.: "Critique of pure reason" (*"Kritiken der reinen Vernunft" (1781/1787)*). Macmillan, Hong Kong, 1929.

Mead G. 1967: "Mind, Self and Society". The University of Chicago Press.

Møller, K.M. 2017a: "Experiencing a painting: An interdisciplinary discussion regarding epistemology and experiencing". Akademisk Kvarter, 16, 65–74.

Møller K. M. 2017b: "Learning and Cognition – The interplay between the Subject and the Group: Understanding the processes of problem-based learning". The Journal of Learning in Higher Education. Vol 13. Issue 2.
Pierce C. 1992:"The Essential pierce. Selected Philosophical Writings. Volume 1 (1867–1893)". Indiana University Press. Bloomington
Pierce C. 1998: "The Essential pierce. Selected Philosophical Writings. Volume 2 (1893–1913)". Indiana University Press. Bloomington.
Reich, W. 2010: "Three Problems of intersubjectivity—And One Solution". American Sociological. Sociological Theory, Vol. 28, No. 1. pp. 40–63.
Weick K.E. & Obstfeld, D. 2005: "Organizing and the Process of Sensemaking". Organization Science, Vol. 16. No. 4. Informs.

CHAPTER 4

Organizational Dialectics and the Becoming of the Organization

Abstract In this chapter we address issues that arise when we introduce a presupposed and formalized set of social rules (organizations) to the social nature of our existence and the contractions which occur not only within the social and formalized rules of organizations but also between the social and formal rules and the nature of human beings. When the organization, in mainstream thinking, is considered as a whole entity in harmony and balance that can be controlled and managed, this is a set of ontological assumptions that logical, historical and empirical cannot be proved in any way as a general argument for organizations and changes. What can be experienced when we see "organizations" is not a clear logical picture of something that everyone can agree upon. It is something else, and a line of questioning rises: What is the organization and the development when humans in organizations counteract the nature of being, for example emotions, perception, identity, self-defining storytelling, and when organizations are humans and their construction and search for meaning, and how should consequences as power and misunderstandings be understood? How can we understand changes and development of the organization and of the human being? A central ontological concept in phenomenology and philosophy is dialectic.

Keywords Dialectic • Contradiction • Subject • Social • Organization • Power • Illusion

K. M. M. Møller, M. Fast, *Investigating Being in Organizations and Leadership*, https://doi.org/10.1007/978-3-030-58138-1_4

In the history of philosophy, the question of how to describe and understand development and changes in nature, history, society, people and organizations has been central and extensive. There are countless situations and historical events that matter in the history of humans and of societies, and for many reasons. The question is how and in which way—how to understand or explain these situations and events, and how to connect these understandings to developments of social contexts and changes of people.

The ontological view in mainstream social science and economics is the idea that a system/whole is seeking balance and harmony, and that the system in itself will find the new balance. The concept of equilibrium is central in mainstream, and is connected to an idea of management as analysis, control and rational decision making based on (complete) information on that whole. The idea that human beings can reach insight in the reality (in itself) that confronts the human being, and as this reality (in itself) can present any knowledge of its own of the organizational reality and development, is both ontological and epistemological questionable. Kant's (1929) theory of cognition and his discussion of *Das Ding an Sich* versus *Das Ding für Uns* shows that an approach to both an ontology of reality and an epistemology demands a perspective of the being of the human being. An attempt to understand development without the human is convicted to fail. There is a need to establish a line of argument from ontology and epistemology of human beings, to development of concepts that can grasp the process of the social and the consciousness, to the action and the interaction of development.

In phenomenology, the concept of dialectic is central in understanding existence and development, and is connected to the human being and social reality in different as well as connected ways. Dialectic can be seen in the theory of cognition, in the view of ontology and in the ground of an empirical method.

Discussions of development and dialectic have been taking place since the ancient Greek philosophers reflected on dialectic in relation to different areas. The word dialectics comes from the Greek *dialegestha* (conversation), *dialegein* (to distinguish/art of conversation) and *dialektikê* (as a method). This focus on dialectics as a method in relation to the dialogue to reach in to something in the other or in reality is central. Of course this is what is seen in the idea of qualitative methods in general—the dialectical idea of understanding as a process. Dialectic was discussed in relation to cognition and knowledge, to method and investigation of something other than the subject and cogito, and to conversation and dialogue. It is discussions on how to understand something and create knowledge (epistemology), and discussions on theorizing of the structure of the world (ontology). A world that is something we are interwoven into, with all of its contradictions and dynamics of reality and of our selves.

The early thinker Heraclitus (around 500 BC) talked of *Panta rhei*—that everything is in movement and all entities move and nothing remains still and the same. He talked of this movement as an ontological view on nature and that objects in nature are in certain position to each other and have an influence on the other—as the flower needs water to grow in the garden. He is known for formulating the statement: "No man ever steps in the same river twice". This has a double meaning, as it can be related both to the movement in nature and also to the perspective that humans change through time passing by and creating experiences, and therefore the human in the river will never be the same human as time and experiences pass. The basic assumption in this is that everything in nature is changing all the time, and that this change is because of a line of contradictions and/or opposite forces in nature or what we face. These contradictions and forces will have consequences as they will relate to and influence each other. As in nature, the changing must also be in and between humans in line with the logic of these contradictions and forces.

Socrates (469–399 BC) talked of dialectic in relation to logical investigation and in connection to the process of dialogue—as a method to investigate someone's beliefs about something: What are the reasons and arguments, and are they logical? He thought (ontologically and epistemologically) that the process in a dialogue of something, and that the process of an investigation was dialectical. That it was to develop something (the truth/a picture of this) by looking into the contradictions of the thing and using language to describe the thing and reach a truth. The idea of conversation as a method and as a process is to reach something new—an understanding or explanation of what is. Plato (428–347 BC) talked about this in his discussion of cognition, and stated that the rational detection of error amounts to finding the proof of the antithesis of situations and events, and to developing a synthesis of what is and was. Aristoteles (385–323 BC) talked about rhetoric and language as being dialectic, and of dialectic as a critical tool in investigation of statements and different meanings of concepts and events.

The tradition of the Greek philosophers and others throughout history leads up to modern times and the current way of thinking. This whole development of thinking can be seen as dialectical, as the discussions between thinkers and ideas also follow the lines of thought of dialectic—that new ideas rise above old ones and above the questioning and disagreements of how to understand and explain. Central to these thoughts was the movement and contradictions, and the ideas to understand them were dialectic, as dialogue, method and thinking. In this way dialectic is important in understanding what organizational being is and the development of organizations, the thinking of people in organizations and the concept of leadership.

Dialectic is at the center of Kant's theory of cognition (1929/1787) and of an understanding of human beings' existence as subjects with consciousness and as subjects in organizations. In Kant's perspective all cognition arises from the experience of something, and knowledge is a synthesis of experiences and concepts—reasoning: for example, without sensing we cannot be aware of any objects (the empirical cognition); without understanding we cannot form an opinion of the object (the a priori cognition):

> There can be no doubt that all our knowledge begins with experience. For how should our faculty of knowledge be awakened into action did not objects affecting our senses partly of themselves produce representations, partly arouse the activity of our understanding to compare these representations, and, by combining or separating them, work up the raw material of the sensible impressions into that knowledge of objects which is entitled experience? In the order of time, therefore, we have no knowledge antecedent to experience, and with experience all our knowledge begins. (Kant 1929/1787:41)

The subject's orientation toward itself and an organizational reality is both the experience of the thing outside the subject and the experience of own thinking, as it is a process of dialectic between what is experienced in the organizational context and what is in the reasoning. It is also a process of imagination of things and naming them. So the process in which knowledge is acquired is composed of sensation, powers of conception and understanding. It is a fundamental conceptual apparatus giving meaning to the world that we are experiencing.

Kant's ontology behind his theory of cognition (as ontology and epistemology are logical dependent of each other, and as such dialectical) is: that the world presented is what we can experience and what we can reason on. Kant distinguishes between the phenomena (the world of phenomena) and reality (the noumenal world), and state in which we cannot apprehend the mysterious substance of the thing, what he called **"das Ding an Sich"**—the thing in itself. This point is central to phenomenology of being, as "the thing in itself" is only reachable through the subject. The thing cannot be reached in any other way—the tree does not present its being by itself. Reason can only be used legitimately in the practical sphere, that is, if we try to acquire knowledge of the world, what we have to be satisfied with is **"das Ding für Uns"** (things as they presents themselves to us[1]). In this way there is the logical contradiction on one hand between the subject's cognition and the world, and on the other, the contradiction between the subject and the social. This can be said to represent the essence of phenomenology, and to formulate the question of what the subject is in its being and how we can reach a social dimension. It is also the ground for the epistemological discussion of how we can create knowledge of the human being and of the social.

This way of thinking consciousness and reasoning can also be found in C S Peirce's (1839–1914) discussion of the research processes as abduction, and indeed is inspired by Kant's theory of cognition as Peirce formulated abduction as a dialectical process of Experience—Reasoning—Hypothesis (a qualitative guess)—Test. The point in his discussion is that object of thoughts is connected to the subject's reasoning and action. This way of thinking can also be seen in J Dewey's (1859–1952): "Learning by Doing" and "Trial and Error"—as they are dialectical processes of learning and developing knowledge and meaning of things and the world. Both Peirce's and Dewey's thoughts are important, as they introduce the process of knowledge as a learning process, and define the human being as a being of learning that exists both in the consciousness and in a social arena.

In this way we have the dialectic of the human being in the world with experience and reasoning, but also with what Kant in the dialectical process of the creation of knowledge thought of as the third point in the process—imagination. The argument for this must rest in the argument that if we did not have imagination, we could only see what we always have seen, and only think what we have thought before. The new thought, the new way of seeing and the creation of knowledge lie in the human consciousness and way of thinking. In this line of argument a perspective on development of the human being and of organizations becomes dialectical

[1] See Husserl's (1962) concept of intentionality.

as a process of thinking, learning and handling the things we encounter in everyday life.

Another dimension in dialectic is the thinking of organizations, history and society, and here Hegel (1770–1831) is important, as he followed the Greek tradition of thinking movement in history as dialectical, as is Kant's concept of dialectic in thinking. Hegel's conceptualization of his thinking of development is the Triadic Structure—that dialectic is movement that can be seen in three stages of development: **Thesis** ("sentence"), **Antithesis**, which contradict the thesis, and **Synthesis** ("composition"), which intercepts what is central as force and as contradictions in movement. Hegel viewed history and the shaping of society as a long dialectical process—it was the forces and different interests that created contradictions, and the movement in society was to address actions toward a solution to them. In this we can see organizations as a social construction that will change as it is not in itself, but rather in the movement of time, caused by the contradiction of power, interests and thinking.

Marx (1818–83) and Engels (1820–95) followed the path of Hegel and formulated dialectical materialism/historical materialism in understanding society and history. In this they formulated three laws of dialectics: (a) the law of the unity and conflict of opposites—it is in this dialectic of the grasping of oppositions in their unity, or of the positive in the negative, that speculative thought consists; (b) the law of the passage of quantitative changes into qualitative changes and (c) the law of the negation of negation (see Hegel), as the form may change over time but the content and the essence remain the same. Dialectic in this perspective is seen as the science of the general laws of nature, the development of society and the movement in thinking, all in line with the overall discussion of dialectic through history, and that the world can in no way be understood as a fixed whole completed all at once. The world is a complex of processes, where things on the surface seem stable but underneath are constantly changing, vanishing and rising, and in different forms over time—the negation of negation.

The Dialectic of Being

Central in the human being as situated in organizations and social contexts is handling contradictions of everyday life and the reality that confronts the subject and subjects. At the same time this process of trying to understand and create meaning from reality is itself dialectical (e.g. Kant's discussion of cognition) and is based on solving the contradictions in our knowing and making sense of situations, things and others. This also relates to what we are discussing in relation to social interaction and the creation of intersubjectivity.

On one hand, the subject is situated with its here and now in everyday life—it is the a priori and is the premise for life and being. We know from our everyday life it is something we have met and that we also take for granted, in general. All of this is grounded in the way our consciousness works, as it is necessary in order to be and to be social beings with others, and to be able to have common activities and common understanding as it arises and is created in this everyday life. The basis for this ability to have something in common is the intersubjectivity in the social origin of knowledge or social inheritance. The dialectic in this lies in the interaction with others and in the subject's ability to transcend the social to become an experience of the subject.

In an intersubjective consciousness (a consciousness about something, not everything), the concept of typification is important as it is the ability to arrange a situation or an object in such a way that it becomes part of a socially important category of situations and objects. People acting toward each other having common typifications are thus enabled to structure their worlds of experiences so that they look alike, through the common meanings created that they lay down on the specific fields of experience important to them. The process of understanding the behavior of others may be understood as a process of typification, where the actors use interpretive structures to understand the object of the actions of other people. These structures come from the experiences of everyday life—"the stock of knowledge" or the "commonsense"—as an understanding lying in the natural attitude of life. It is through the use of typification that we classify and organize our everyday life reality.

Typification can be understood as a two-pronged process: on one hand, it solidifies some aspects of reality into an incorrigible, self-evident field of "un-problems"; on the other hand, it draws the line that renders everything left on the other side potentially problematic. That is to say, the process of typification determines, by the same token, what is to be determined and what is to remain indeterminate. The stock of information includes the knowledge that other people like us exist in the world and that their conduct has the same structure which we "know" from the experiences of our own acting in everyday life. This knowledge renders other people potential partners in communication when viewed as a "trade in meanings", and as a mutual effort to grasp the message conveyed by words, gestures, facial expressions and so on. Other people differ from all other types and from inanimate objects in particular. In short, people's conduct is to be interpreted as a basic voluntary and purpose-oriented action (Bauman 1978:184).

Everyday life is based on the certainty that objects and events can be interpreted in the way that we are used to. That the world exists in the way that we take for granted and has not changed since last time—that we have the interpretations, meanings and language to make sense of it. In the natural attitude toward life, we have an underlying anxiety that it could be other ways than what we think. But the broader picture is what we trust in. We are not ready to leave this natural attitude toward it without having experienced a certain shock that forced us to break through the boundaries of this finite province of meaning and change the accent of reality to another. These experiences of shock are often made in the midst of everyday life. They are part of its reality themselves. All of this is a part of life and being human, and grasps the dialectics of just this being. Humans develop in meeting something that is different, something we do not know, but something we need to handle and gain an understanding of. This understanding as a process (as we discussed in Chap. 2) is a dialectical process, that understanding of something also needs to imply an understanding of oneself and own thinking—cogito of cogito. It also implies that central to being is knowing, making sense and having the here and now as something certain in life. Anxiety is a state of mind and is something that needs to be handled, either constructively by solving it or less constructively by fleeing from it and hiding.

The Dialectic of Organization

The idea of an organization is in itself dialectical, as it is both an illusion and something very concrete, as it appears in an arena where we meet people face-to-face. The illusion and the concrete meet in the organizational context, and will support each other or will appear as contradictions.

The organization is at the same time both stable and an illusion; it is social and it is a discourse of repression and alienation; it is a possibility for emancipation of the potential of the human and it is a force of power. The dialectic of this is that the organization both is and is becoming. The organization, like any other phenomenon, is from a perspective and a position.

The idea that the organization, in the ontology of rationalism and positivism, is in itself a whole, seeking balance and harmony, is the illusion. Organizations are everything else and are full of contradictions, especially when we reflect upon them in the light of time.

> The word organization is a noun, and also a myth. If you look for an organization you won't find it. What you will find is that there are events, linked together, that transpire within concrete walls and these sequences, their pathways, and their timing are the forms we erroneously make into substance when we talk about an organization. (Weick 1979:88)

> When we look at organizations, especially the larger, older, famous ones, they seem *solid*, they seem *permanent*, they seem *orderly*. This is, after all, why we call them organization. Images of organizations as solid, permanent, orderly entities run through many textbooks. But, in our view, they only tell half the story. They obscure the other half: the chaos which looms behind the order, surfacing from time to time, such as when computer systems break down, when products are sent to the wrong destination or when bookings are made for the wrong dates. They also obscure the immense human efforts and energies which go into keeping organizations solid and orderly. (Sims et al. 1993:1)

Organization is a concept and as such always refers to a position and a perspective, and also therefore in the related language. The word *Organization* says nothing, as it is the meaning of the word that says something. This meaning is, as we have discussed, the ongoing result of a dialectical process. The dialectic of the organization appears in the language used in the organization, as it can disappear when the language in the organization and the perspective in play are no longer enough to form an understanding and create meanings out of situations and contradictions. The Organization is a conceptualization of what we believe in, and also what we take departure from when we orientate our actions toward things. Organization is a concept used as a typification of an arrangement of experiences (e.g. company, department, family, school class, union), and is as such a part of the tradition and the history that is the foundation for just this typification. In this sense it is an illusion as it only exists as long as people believe in it and act in relation to this belief. The confirmation of the organization is in the belief, as it is real in the sense that it works for people and their everyday actions, and in the language, as the language confirms that the organization is what people experience and can understand and agree upon organizational settings and discourses. In this way there is a dialectical logic between the illusion and the action, where the action confirms that the illusion is real and can be trusted.

The organizational tradition and histories are created by this dialectic process and constitute something that is social, and moreover, the social becomes more than the beliefs. It becomes a manifestation of the beliefs and actions. It materializes, becoming physical with buildings and interior design, and it is told stories from the history of the organization. The stories are ongoing and also contradict each others, and are exposed for attempts to be controlled by different groups. But some become the ruling stories connected to the major discourse of the organization and can be legitimized by society. The management will act according to their story and will try to legitimate their position by repeating that story. Different groups of interests will be supported, and as long as there is a legitimation of the organization, people will keep believing in the organization and act in relation to that. The process in this becomes dialectical as there is the illusion vs. action connection, and contradictions between groups and power. What will be the fundamental premise of the possible existence of the organization is that it has the legitimation in the ruling discourse in that society.

Organizations are in our life and are the part of the socialization that makes people become organizational citizens. They are a natural part of our lives and are taken for granted, also because they work for people or, more accurately, people make them work. The organization becomes a collective arrangement where people give situations and activities meaning. In line with Blumer (1986), organizations consist of *the fitting together of lines of activity—the interlinking of lines of action*. These fittings are action parts as they arise in intersubjectivity and in that which people take for granted. We actively construct our social reality through language, through a process of symbolization by forming words and sentences to describe our experiences as well as our wants and desires. We create our organizational existence in a dialectical manner: "A dialectical view is fundamentally committed to the concept of process. The social world is in a continuous state of becoming – social arrangements which seem fixed and permanent are temporary patterns and any observed social pattern are regarded as one among many possibilities" (Benson 1977:3).

An Organization is not one thing. It is not one system, not one whole. The organization is a complex multi-level phenomenon, where different interests and ideas are thought and acted by people, and they are at the same time both involved in social formations and at a distance from each other in spite of their different positions of power and symbolic capital. It is this that creates the movement and the becoming of the organization. The dialectical processes are in the production, reproduction and destruction of organizational forms and discourse. At the same time it is a perspective of understanding people as creators of the rationality and stability that are needed to maintain the organization as something stable in the organizational discourse.

Actors have different motives and definitions of situations, but they also have some in common that make the social world of the organization into an inner logic, with rules and lines of action deriving from situations and the history of experiences. On one hand the organization is a stable social formation on the surface level, but on the other hand the social depends on people embracing it as a collective idea that works in the everyday and make sense to most people. The actors involved in the organization expect suitable actions from themselves and from others: they are capable of understanding meanings of action by others and having their own point of view about themselves based on the response of other actors. They associate meanings with situations and with other actors' actions and act in relation to their interpretations of these meanings. This can be understood in relation to typifications, formed by earlier experiences, which define humans' *thinking-in-future* about others' possible reactions to the subject's actions. The problem arises when the inner logic of the organization is exposed to contradictions and becomes a power game and/or is exposed for something unfamiliar, and the situation will demand a new inner logic in line with Hegel's dialectical thinking.

This social reality in the organization is pre-defined in the language by which members are socialized, and creates that inner logic. There is a language in play with meanings and discourse. The language defines and emphasizes some of the experiences of this social reality. The language spoken and the dialogues between actors in the organization form this inner logic and maintain the organization as the idea of something concrete and familiar. But in situations where actors meet something new or the social reality is characterized by power and the rule of the game—the doxa—changes, the development of a new language is needed. Language is the means to create a new understanding, changes in meanings and a new worldview. Language is the baseline from which we understand and can interpret knowledge. The process of the new language is also dialectical, as the established language was created by social processes of meanings by the subject, and the new language needed has to become the new language through dialectical processes handling the contradictions and creating new words and concepts to do this.

The focus in understanding the organization is the way organizational members interpret their organizational world, which is essentially a special sphere of the individual's Lifeworld. Lifeworld refers to the fact that in any life experience there is something that is given in advance or something that exists in advance and is thus taken for granted. This taken-for-granted world includes our everyday life and whatever prejudices and typical interpretations we may derive from it. The important characteristic of this experience in any organization becomes the typical form of everyday life. Or as described by Schutz (1990:7): "The individuals commonsense knowledge of the world is a system of constructs of its typicality". In social interaction, the role of typification is important and can be expected to vary according to the nature of the relationship.

The Context

The space and time of the context means that something is close and concrete, and is where the actor travels and interacts. This can be seen in the consciousness of human beings in "the natural attitude" first of all being interested in that part of everyday life that is in the subject's reach and that in time and space are centered around the subject (see Schutz 1973:73). The place where the body occupies the world, the actual here, is the point from which one orientates oneself in space. In relation to this place where one is situated, one organizes elements in the environment. Similarly, the actual now is the origin of all the time perspectives under which one organizes events in the world as before and after, and so on. This space is experienced by the actor as the core of reality, as the world within reach. It is the reality in which we are all engaged and is in our being.

Any development can be seen in relation to the actor's everyday experiences with the attempt to orient and to solve problems. When the actors act in their space, they thus widen their understanding of reality by interpreting and relating themselves to the result of the actions. In this way the dialectics of understanding will create the knowledge by reflecting upon the experiences and the context, as the knowledge will be the context.

Development of knowledge is cognition, as discussed in Kant's (1929) dialectical theory of cognition. It is, in everyday life processes, interpretation and retrospection whereby the subjects create their space: reality is what one sees; hence it changes every time the actor constructs a new concept and language of connections and experiences.

The idea in this is that we develop knowledge through actions and that actions are the means by which we engage ourselves in reality; our actions construct and keep us in touch with the world and keep us in touch with ourselves (Garfinkel 1967; Morgan and Ramirez 1984). The action-knowledge discussion is built upon the assumption that we only have a reality in force regarding that with which we are engaged: reality is constructed by the subject and is also socially constructed. This engagement and the interpretations, and the fact that we are situated in our here and now, also imply that people have some control over the process of constructing reality or that they have opportunities to change it, because they do not act alone and because it is an ongoing process.

Interaction as the Dialectical Ground

Interaction is both concrete in its movement and is symbolic in the sense that actors respond to the actions of others, not for some inherent quality in them, but for the significance and meanings imputed to them by the actors trying to read them. Meanings shared in this intersubjective way form the basis for human social organization (Singelmann 1972:415). People learn symbols through communication with other people, and therefore symbols can be thought of as shared meanings of something (Rose 1962:5). This mutually shared character lies in intersubjectivity and stresses that it is interaction that constitutes any social formation. The content relates to the meanings, and therefore the content in a social logic is in the intersubjective and a process of the dialectics of understanding each other. It is, however, a volatile logic, as it depends on the retention of the interpretations.

Each interaction implies possibilities for experiences and information, and for strengths or weaknesses in interpretations of connections in the situation. In every situation there is the possibility of several different interpretations. This means that changes in the experience of the context create ambiguity and the actors are tempted to use previous successful actions and interpretations.

An organization is overlapping interactions, where the subjects create the organization as meanings and as ground for actions. But "it" (as the illusion) also has an influence upon them through their interpretation of "it". This dialectical perspective appears from the view that the organization only exists through the interactions between the subjects and is thus viewed as a corollary of these interactions. Simultaneously, the organization is historical to the individual member: the individual enters into an already existing everyday organizational life, which sets the possibilities or restrictions for the subject's self-development. Self and organization thus develop together and because of each other in a dialectical process of mutual transformation (Singelmann 1972:415; see also Mead 1962; Berger and Luckmann 1966; Benson 1977).

The actors have to live with and exist with uncertainty and ambiguity. In other words, the way in which the actors handle themselves is in itself uncertain and exposed to different interpretations and possible consequences. To reach some kind of security, the actors attempt to organize their activities and make sense of their interpretations. In this organizing the dependent actions are oriented toward removing contradictions and uncertainty: the actors seek to define and make sense of their situation, and thus they create both the firm and the experiential space. Organizing is to be seen as a social, meaning-making process where order and disorder are in constant tension with each another, and where unpredictability is shaped and "managed". The raw materials of organizing are people, their beliefs, actions and shared meanings that are in constant motion (see Sims et al. 1993:9).

Actors create a reality that contains specific actions (routines, traditions, procedures, politics, myths etc.), mental maps of the context, norms and values (as symbols). All of this relates to the interpretations and expectations that make sense to the individual subject and to the subjects in a social context. In the social subsystem constituted by a formal organization, the assignment of meanings is not left to the discretion of the members alone. The organization presents the individual member with a number of anonymous, functional schemes that will help the subject to orientate behavior toward the incumbents of other positions, especially hierarchical positions and power. These confer on the newcomer in the organizational hierarchy organizational titles and formal areas of responsibility. They underlie job descriptions, the rights and duties attached to each organizational position, rules of conduct, customs and so on. Through such standardization, the organization attempts to establish a congruency between the scheme used by the actor as a scheme of orientation and that of organizational colleagues as a scheme of interpretation. This standardization is supposed to promote the smooth flow of authority relationships required for the efficient functioning of the organization (see

Jehenson 1978:226; Benson 1977). The situation accordingly is that reality is never the same as the schema, and contradictions will arise when these breaks down or are misinterpreted, when people do not care, or when some think they are correct and act according to that logic, and others do not. Schema and standardization can work when the situation is simple and can be repeated. Most of life does not run in this way.

The organizing is also in relation to the extent of the actors' involvement in an effort to retain or to change the rules, or their attitudes toward them. The actors solve problems with developed definitions and take actions in relation to the dominant views of reality. This is the ruling discourse and will always be in the view of the power. Actors act rationally and logically according to their understanding and interpretation, or in relation to the extent that fear is present—either directly as in a situation where people are fired, or in a situation where changes might be dramatic for people in the organizational context. Acting rationally is a social construction and a social product rather than action guiding rules for organizational life. It is also a symbolic product, constructed by actions dependent on the actors' interpretation of reality and required interaction. That is, the structuring of organizational interactions requires members to rely upon shared but largely tacit background knowledge that is embodied in an organizational discourse or paradigm (Brown 1978:374; Garfinkel 1967). Rationality as well as the definition of "problems", "situation", "leadership" and so on are afforded by the dominant discourse, and by that defined as the truth.

As long as organizational actors act as typical members, they tend to take the official system of typification for granted, as well as the accompanying set of recipes that help them define their situation in an organizationally approved way. The emergence of other, non-organizationally defined schemes results from the breaking down of the taken-for-granted world when the actors enter into face-to-face relationships. The force of power also creates the social norm for acting, and this is not the same as people acting from a moral foundation. People act to maintain life as they know it.

The Organizational Dialectic in the Totality

The organizational context and the becoming lie in what the actors interpret as their situation, market and surroundings, understood in relation to situations where the actors interact. But the context is also the social situations and the relations that the actor has with other actors "outside" the formal organization. This is understood as a broad spectrum of relations and involvements in interaction connected with knowledge and understanding of reality such as friendship and family, or seeing reality as a "multiple reality". In other words, the actors do not only create their understanding of reality at work between 9:00 and 5:00. They have a history and a life aside from their involvement in the firm. This is of course the point of this discussion, that the being is not the same as the function or the role in the organization. People are much more and something else.

The organizational context is the actors, and it is the totality, as understood with all the complexity of the society, ruling discourse and competing discourses. The dialectic is not only between the subjects in an organizational context and situation. It is also what meets the subjects in professional life, in family life and every other social context. It is what people are confronted with in society, and that they define as something they need to handle. There are all the regulations and laws from government that have their dialectic in development of society and organizations. All of these are a part of the general discourse and also the ideas of norms and values in society.

An example of dialectic and development can be seen at the micro level, when the subjects interact with a stranger. The subject enters the process of interpretation and forms an understanding of this stranger's view of reality and how it is interpreted. If we talk of a stranger in a community there will be a need to construct intersubjectivity, which is the foundation for a further interaction. Their approach is comparable to a researcher's approach in accordance with qualitative methodology and the dialectic in Plato: the approach by which we understand others' actions and the meaning of those actions.[2]

The understanding of action and knowledge processes in the organization can be discussed in a dialectical frame based on three areas (see Clark and Fast 2019): the actor's development capability, organizing and the inner logic, and the actor's extension of the context.

The Actors' Development Capability can be seen as a process of actions. The capability in this is in interaction, but where the state of development is dependent on the interpretations and the actors' capacity to create interaction and understanding of these. The actors, with their

[2]See Weber (1948, 1972); Goffman (1959); Garfinkel (1967); Giddens (1976); Schutz (1978, 1982); and Morgan (1983).

specific qualifications, experiences and personalities, are not passive participants in a prefabricated reality; they are contributors and creators of meanings (see Allaire and Firsirotu 1984). In this fabrication the becoming is produced and is the movement as a dialectical process.

The interaction is an involvement of something in a situation. But the process of development does not start with the situation or the involvement but with a combination of both, based on the dialectics of interpretation and expectations of the future and in processes of interaction. The organizational development can in this be formulated as the actor's development and readjustment capacity to change understanding and act in a new way. This can be seen as the ability to develop knowledge and transform this into actions, realized through the process of interaction in the organizing.

Organizing and the inner logic can be seen in the actors' understanding of formal structures, goals, politics, management processes and resources, which is a result of the organizational way of functioning as based on agreements, typifications and what is taken for granted in everyday organizational life. It contains a living dimension in a connection of shared symbols manifested in interactions, interpretations, myths, ideology, values and multiple cultural forms of expression as the result of interpretations and lived experiences. An important part of this everyday life is the business area, which is based upon previous actions oriented toward what to produce, which resources and technology one should use and an understanding of the existing knowledge and needs for knowledge development in relation to the business area. This could also be understood as what Diamond (1990:34) sees as "the organizational identity". Organizational identity is the product of the group's intersubjective organization of experience at a given point in time: the "story" they share about what is real to them. It is a picture of the meaning, purpose and intention collectively and unconsciously assigned to common experience and behavior of organizational members, especially during critical incidents. Or what Benson (1977) calls "organizational morphology", which refers to the officially enforced and conventionally accepted view of the organization. It refers to the organization as abstracted from its concrete, intricate relations with other aspects of social life. This is the administrators' vision of the organization, the form that they try to impose upon events. Since they are partly successful, the morphology may also be somewhat accurate as a description of the organization.

It is the connection between the actors' knowledge, their understanding of the business area and their organizational actions that creates the orientation toward and the mode of handling organizational activities. This should be seen in the light of the actors' attitudes and intentions toward the activities. The actors' development ability is constituted through the interaction process and by the actors' interpretation and knowledge of the business area. The actors' change of perspective is therefore important in a situation where contradictions exist between the business area and the experiences of the context and situation. This can be seen in situations where one shifts to new technology, new products, entering a new market and changing services. In each of the situations there is a need to change knowledge. Therefore a recognition among the actors of the problem, new actions or actions in a new way is important, and they need to enter a process of dialectic in different dimensions: the method to investigate and create experience, the cognition process of the subject and the development of a common understanding, and the creation of a new everyday life, and a language to handle and describe all of this.

Through the interaction process and interpretation the actors' experience of context influences the organizing and the development capacity. There is a confrontation with other actors' external in situations, and when the situation is experienced as uncertainty or ambiguity they will need to enter into the dialectic of understanding the context and themselves in another way. The foundation for any experiences and interpretation, for development of knowledge and for changes in organizational activities, is the actors' interaction and involvement, and self-development. In situations with changes, the central issue is the actors' knowledge and change of interpretations that can be related to and transformed into actions.

As discussed in this chapter, the understanding of movement and development of organizations is closely linked to the concept of dialectic. The dialectic exists in the human as cognition and understanding of oneself, and is in the development of the organization as the handle of the contradictions that arise. All of this raises the question of how to understand the leadership in dialectical organizations.

References

Allaire, Y & Firsirotu, M E:" Theories of Organizational Culture". Organizational Studies, Issue 1, no. 5, 1984.

Bauman, Z.: "Hermeneutics and social sciences – Approaches to understanding". Hutchinson & Co, London, 1978.

Benson, J K: "Organizations: A dialectical view". Administrative Science Quarterly, mars 1977, vol 22.

Berger, P L & Luckmann, T: "The Social Construction of Reality – A treatise in the sociology of knowledge". Doubleday & Company, New York, 1966.

Blumer, H.: "Symbolic Interaction – perspective and method". Englewood Cliffs, N.J.: Prentice-Hall, 1969, 1986.

Brown, R H: "Bureaucracy as praxis: Towards a political phenomenology of formal organizations". Administrative Science Quarterly, sept. 1978.

Clark, W Woodrow II and Fast, Michael: "Qualitative Economics – The science of Economics", Springer, Cham Switzerland, 2019.

Diamond, M A.: "Psychoanalytical Phenomenology and Organizational Analysis". Administrative Science Quarterly, Spring 1990.

Giddens, A.: "Hermeneutics, ethnomethodology, and problems of interpretative analysis". In L A Coser & O N Larsen (eds): "The uses of controversy in sociology", The Free Press, New York, 1976.

Garfinkel, H: "Studies in Ethnomethodology". Prentice-Hall, Englewood Cliffs., 1967.

Goffman, E: "The Presentation of Self in Everyday Life". Doubleday, New York, 1959.

Husserl, E.: "Ideas". Macmillan, New York, 1962.

Jehenson, R: "A Phenomenological Approach to the Study of the Formal Organization". In Psathas G. (ed): Phenomenological Sociology – Issues and Applications. John Wiley & Sons, New York, 1978.

Kant, I.: "Critique of pure reason" (*"Kritiken der reinen Vernunft" (1781/1787)*). Macmillan, Hong Kong 1929.

Mead, G H.: "Mind, Self, & Society – from the standpoint of a Social Behaviorist". The University of Chicago Press, Chicago, 1962 (1934)

Morgan, G (ed): "Beyond Method – Strategies for Social Research". Sage Publications, Beverly Hills, 1983.

Morgan, G & Ramirez, R: "Action learning: a Holographic metaphor for guiding social change". Human Relation, vol 37, no 1, 1984.

Rose, A M: "A systematic summery of symbolic interaction theory". In Rose, A (ed.): "Human behavior and social processes – An interactionist approach". Routledge & Kegan, Paul, London, 1962.

Schutz, A: "Collected Papers I: The problem of social reality". Kluwer Academic Publishers, Dordrecht 1990.

Schutz, A: "Hverdagslivets sociologi". Hans Reitzel, København 1973.

Schutz, A: "Life forms and meaning structure". Routledge & Kegan Paul, London, 1982.

Schutz, A: "The theory of social action". Indiana University Press, Bloomington/ London, 1978.

Sims, D, Fineman, S & Gabriel, Y: "Organizing "Organisations"". Sage Publications, Great Britain, 1993.
Singelmann, P: "Exchange as symbolic interaction: Convergences between two theoretical perspectives". American Sociological Review, vol 37 (Aug.), 1972.
Weber, M: "The Methodology of the Social Sciences". The Free Press, New York, 1948.
Weber, M: "The interpretation of social reality". Thomas Nelson & Sons Ltd, London, 1972.
Weick, KE: "Cognitive processes in organizations". In: "Research in Organizational behavior", vol 1, 1979.

CHAPTER 5

Leadership as the Direction of Being

Abstract Throughout this book the analysis of being has provided an existential description of what it means to be and to be an individual in an organization. Regarding popularity in literature and in a general discourse in society, none are more so than those individuals at the top of the organizational hierarchy, often referred to as managers or leaders. This chapter will distinguish between managers and leaders due to the qualitative difference between the characteristics of the two roles. The two concepts can be defined in various ways but since the goal is not an etymological discussion, a brief definition of the two concepts can be presented: a manager organizes and distributes resources, as opposed to a leader who is highly significant in setting the cultural and/or practical direction of a group of people. This chapter will address the notion of leadership from an existential phenomenological perspective with the goal of providing contextual and cultural discussions regarding what leadership is and potentially can become. Leadership is missing in organizations today and it is substituted by managers who without explicit moral aspirations such as those found in leadership can without analysis accept and support ethically and culturally problematic assumptions regarding the human being.

Keywords Leadership • Knowing oneself • Knowing the others • The good • Moral • Self-critique • Reflection

K. M. M. Møller, M. Fast, *Investigating Being in Organizations and Leadership*, https://doi.org/10.1007/978-3-030-58138-1_5

Throughout this book the analysis of being has provided an existential description of what it means to be and to be an individual in an organization. There has been made no distinction between the different functions and roles which different individuals might occupy in organizations. Regarding popularity in literature and general discourse in society, none are more so than those at the top of the organizational hierarchy, often referred to as managers or leaders. The discussion will distinguish between managers and leaders due to the qualitative difference between the characteristics of the two roles. The two concepts can be defined in various ways but since the goal is not an etymological discussion, a brief definition of the two concepts can be presented:

- A manager organizes and distributes resources.
- A leader is highly significant in setting the cultural and/or practical direction of a group of people.

Kirkeby (2008:9) states that "Managers have stopped being leaders". This remark is here interpreted as a provocative comment on the current organizational development and discourse. When looking at some organizations, we are inclined to agree both with the statement and with its critical implications for organizations and their members. Due to this organizational critique, this chapter will address only the notion of leadership from an existential phenomenological perspective with the goal of providing discussions regarding what leadership is and potentially can become.

Regardless of perspectives, any definition of leadership has to overcome the challenge of explaining how someone can get other people to act in a certain way and do what the person wants them to do. Now, we might experience that sometimes we can convince other people to do something for us, or, even more manipulatively, sometimes we can get them to do something which is not to their benefit, but we get them to do it anyway. Such an experience might lure the manipulator into inferring causality between one's persuasive abilities and the outcome of the other's actions. From a phenomenological perspective, the notion of persuading someone to do something is problematic as it implies some measure of direct influence upon the decision making of the other person. The will of an individual might not be absolutely free and thereby cannot be understood as a completely spontaneous phenomenon. Instead will and decision making can be understood as the notion of interpreting and acting in relation to a given context within a historical narrative. The idea of making a decision is therefore as much a matter of identifying the context and its potential as it is about deciding what is desirable in that context. Those two notions, identifying the context and knowing what is desirable, both imply that the leader is capable of performing such mental tasks in a way that is sufficient for his given context. Leadership is therefore about knowing oneself, knowing the others and knowing the evaluative criteria for a given context. Kirkeby expands on this:

> *The leader, then, holds power, and therefore enjoys a position of justified authority; first, because he possesses insight into the structure of the organization, its principles, its values; second, because he can construe these in the light of the survival of the whole; and, third because he has the ability to transform the goals of the community.* (Kirkeby 2008:19)

Because the leader holds a position of authority it is necessary that the position is justified, which refers to a culturally valid argument regarding the position of a leader and the leader. This justification is significant both in regard to the employees and, if they exist, the people who lead the leader. This notion of justification highlights that leadership is dialectical, and cultural, resting upon the empathetic ability of being the leader.

A second element Kirkeby mentions is the leader's ability to transform the goals of the community. Either this implies a notion of control over the decision making of the employees, which we doubt Kirkeby intended, or it implies a capacity to significantly influence the intersubjective elements within the group of which the leader is a member. In his definition of leadership, Souba is more specific as to how this influence might be performed:

> *Being a leader means using language to reframe people's challenges such that more effective ways of being and acting are the outcome.* (Souba 2014:80)

A significant characteristic of a leader therefore has to be a certain and significant ability to analyze and take part in groups in a productive and developing fashion. Both Souba and Kirkeby place learning and development as central goals for leaders and subsequently ideals, which are significant for their employees as well. In Souba's description the function of leaders is to create more effective employees through reframing. Unless the notion of effectiveness is in a significant manner shaped by the perspective of the employees this appears rather unethical (because then the top management has hired a leader to get people to do what they want them to do, and not necessarily what is best for the employees). When effectiveness is linked to a narrowed economic rationale this will be a problem for the leader, the company and the idea of being effective in the long run. Throughout history, we have seen countless examples of work that is harmful to employees. The reason for the criteria of effectiveness is abundantly clear in modern societies, but should not be the direct measurement for the behavior of others or employees. Effectiveness tends to hold a very shortsighted notion of performance measured through specific criteria above all other facets of life.

Kirkeby presents a broader perspective of the reframing element of leadership when he states that "leadership consists in the will and ability to form itself in the image of the good" (Kirkeby 2008:45). The good has been defined in many ways throughout history, but for this discussion the concept describes a contextual ideal in line with the golden rule (*do unto others as you would have them do unto you*) and a concept of human beings. We know the Good from the Golden Rule, that is, from the experience we draw from our social contexts, where we identify ourselves with others. "We can let norms that either make too little of the Golden Rule or too much of the "business at hand" take control of our inner dialogue, our soul's conversation with itself" (Kirkeby 2008:25).

The ladder in this book is defined as the existential functioning of human beings, and the knowledge of what it is *to be human* is central to leadership as it provides a stable set of tools for introspective understanding as well as the understanding of others. It keeps the leader from being swept away by the cultural and often implicit ontological winds of contexts.

In a concluding fashion, Kirkeby describes leadership as follows:

> …. *Leadership is only possible on the basis of a clear concept of self-leadership that can maintain what is distinctive about leadership across all situations and every dynamic context. This problem is at bottom pedagogical but in a broader and more fundamental sense than we otherwise are used to using this concept. The leader should be precisely a pedagogue, both for himself and for others. The term "pedagogue", which originates from the Greek ideal of education, namely, "paideia", doesn't only allude to the teaching of children, but, more precisely, to lifelong process of learning and the development of a socially relevant formative process.* (Kirkeby 2008:45)

This description places a phenomenological understanding of concepts of learning, interaction and ethics as central to the lifeworld of the leader. However, where there might be an issue is in the implicit notion of the leader as the initiator and the somewhat unaffected individual. *The leader should be a pedagogue, both for himself and for others* is a statement which to a large extent is accurate in terms of the responsibility of a leader, but it does appear as if the leader somehow transcends his context of governance. Due to the minor notion of this comment it is unclear as to whether either Souba or Kirkeby would disagree, but it is our perception that both authors to a degree infuse the leader with an ability to navigate somewhat unaffected by the context in which the leader exists. Leadership in our version of a phenomenological perspective is co-existing and co-creation; it is dialectical and reciprocal. Therefore the notions of *reframing people's challenges* and having *a clear concept of self-leadership that can maintain what is distinctive about leadership across all situations and every dynamic context* are beneficial ideals but seem to fall somewhat short of understanding and emphasizing the complexity of co-existing as a leader. The context and the inevitable role it plays in our perception also affect the role of the leader. Leadership and its reference to justification and the good across contexts are therefore understood as ideal goals which are co-constructed, and the reframing aspect of leadership might more aptly be evaluated and defined in accordance with the context, historicity and existential functioning of the relevant leader and employees. Such a relationship is also evident in Johnsen's (2002) discussion of the entangled relationship between coworkers and the leader. It might appear hypocritical that we argue against the stability of leadership when we argue for existential functioning as the stable element of leadership, and this deserves some clarification. We have previously shown that being a human being includes a set of existential functions that are relatively stable over time (although affected by the content of the functions, e.g. our perceived everyday lives). Existential functioning is therefore central to the possibility and foundation of leadership. The content of existential functioning, in this case the historicity and context of a given leader and the correspondingly enacted leadership, therefore does not transcend contexts any more than the rest of our lives do. Leaders and everyone else are left with the impossible task of completely understanding themselves and their relationships as they unfold in a dialectical and continuous existence where only retrospective sense making offers some comfort. This paints a picture of leaders as individuals, for whom self-analysis is a significant part of themselves and their job description. The leader must create the self each day and in every moment, but self-creation is tricky as it is complicated and comes with contradictions. This self-creation is not the pure development of natural talents but a process that only makes sense in a particular social context (Kirkeby 2008:29). So in this we have the dialectic between the leader and those being led in a context, and there is self-creation as the process of being aware of oneself and being aware of something. To be conscious of oneself in understanding is always at the same time self-understanding—the cogito cogito—to be conscious of your own consciousness.

While reflexivity and analysis are certainly relevant for leaders, the retrospective nature of reflection lends itself very poorly to the present. Being a leader is also about being in the present, where reflection is impossible as it quickly draw us away from the present. The attempt to be in the present is a process, in which I am continuously giving myself the task of being present in my life and therefore of being present for and with others (Kirkeby 2008:122). Being in the present and one's immediate actions implies that the interactions can be guided not by conscious consideration but by something else. Immediate being implies a complete directedness toward someone else. Here the previously mentioned concepts of I-Thou and I-It can shed some light on the ways in which the intentionality of humans can take shape and direct itself toward someone else. The two relationships among other things describe the possibility of the directions and intensity one's intentionality and being can take. Forgetting oneself in I-Thou relationships is central to understanding the essence of such situations. Self-forgetting includes forgetting you are a leader; it is simply about being there as you are as a person. Who you are as a person is naturally a product of which narratives you have and what interpretation of the given context you have. Therefore, ideas of specific types of personality that are more suited for leadership are understandable but logically problematic since it is impossible to completely take into account the dialectical relationship between leader and context and how that will unfold. Here there exist three dilemmas for leadership in practice, which we will discuss further.

The first dilemma is that by being in I-It relationships the leader does not follow the golden rule since it is fair to assume that no one wants to be treated only in a strategical and calculative manner. This is in line with Kirkeby's notion that leaders need to justify their position since their very position is one that does not follow the golden rule and therefore they must justify themselves by knowledge and with clear ambitions to follow the good and what benefits the organization as a whole (Kirkeby 2008:19). In other words, because a significant part of the leadership role is assuming an interpretational position which is calculative beyond what is otherwise acceptable, for example remaining in the I-It relationship in order to structure and influence social aspects of other people, the leader must justify the leadership position taken by referring to the good (which to some degree is contradicted by remaining in the I-It relationship). This dilemma appears unavoidable and necessary for the leadership role, and only to a degree can be justified by aspirations toward the good.

The second dilemma is that the leader is, due to the position and function, supposed to strategically direct the actions of others in order to achieve a goal. This means emphasizing the goal rather than the means toward that goal, and in turn it implies treating others as a means to an end. It means intentionally remaining in the I-It relationship in order to perform their job. Doing so will, as Buber points out, develop a numbness toward I-thou relationships, which are crucial for creating a community and personal relationships with people. It is precisely the point of emotions and inspiration that seems significant for the original distinction between managers and leaders. As shown in previous chapters, leaders as well as all other individuals are beings, whose actions stem from an emotional existence. Emotionless rationality is a conceptual illusion due to the nature of our intentionality and interpretation of reality and of ourselves. When individuals attempt to act in a leader-like fashion toward others they can become leaders only to the extent that the other individuals choose to follow. That choice is, like all others, based on an emotional response toward the perceived context and individuals. The existential functions of human beings and the nature of the corresponding intersubjectivity are central to understanding the choices and interpretations of humans and therefore also regarding whether or not to follow another individual. In other words, leadership is an ability which not only rests on reflexivity but is also substantially dependent upon the emotional insight and understanding of both the leader and those being led. As such, leadership appears to be a dialectical and intersubjective phenomenon since it is not possible to lead without there being someone willing to follow (this applies to self-leadership as well). This emotional codependence is central to the notion of leadership. However, it stands in opposition to the notion of managers as managing resources or leadership and strategically directing employees due to the nature of I-It approaches. The second dilemma therefore requires a balance between remaining in the I-It relationship for coordination and striving for the I-Thou relationship for understanding and familiarity with the employees.

> *The primary word I-Thou is not of evil – as matter is not of evil. It is of evil – as matter is, which presumes to have the quality of present being. If a man lets it have the mastery, the continually growing world if It overruns him and robs him of the reality of his own I, till the incubus over him and the ghost within him whispers to one another the confession of their non-salvation.* (Buber 1958:41)

Throughout the work of Buber it is established that the I-It relationship has an effect on the ability of an individual to enter into I-Thou relationships. The ghost and the incubus whisper their non-salvation as a response to the world of It, an aspect of reality which manifests itself in many fashions in organizational life. As the world of It stands in opposition to existential engagement, self-forgetting and immersive experience, we argue that the quantification of everyday life in its many forms encourages a life lived in the world of It. A few examples are performance-based management, functionalistic management and social networks. The world of It is necessary as it allows planning and it is beneficial in several ways; however, it creates problems with the openness toward immersive experiences and therefore with the very ability to lead. The lack of I-Thou is problematic in itself, but certainly more so due to the responsibility and capacity of a leader.

The third dilemma can be derived from the previous quotation by Buber and it is the notion of the growing world overrunning the individual if it manages to claim the quality of present being. Here lies a significant distinction which is enabled through the dialectical thought process and challenged by our own volitions and the discourses of organizations and society. The dilemma involves pursuing the good in an autonomous fashion and at the same time adapting to relevant perspectives presented by others. The aspiration of good is normative and contextual, which means it entirely depends on the ability to select and de-select elements in the different perspectives and discourses the leader might be exposed to. Therefore immersing oneself in the relationship with others and navigating in organizational contexts imply not only co-existence, but also a significant understanding of the different perspectives, positions of power and so on that create and change the organization.

Leadership is being human, and is also being humanist in many ways. It is in the will and in the ability to form oneself in the image of the good. Leadership is dialectical as it embraces the process of oneself as a reflective being, and also the other in goodwill and in an authentic interest in the other. Leadership needs to be dynamic, as the dialectical perspective points out, but also in the development of the self, which is a learning process. The learning must also be in the relation to the other in a pedagogical relation. Not only as some form of coach, but also the other way around: a leader needs to learn from the other. This discussion of the leader is complex as the leader is situated in the organizational context with a formal position that possesses power, and the relations to the other will always be unequal. The matter of power, goodwill and legitimacy can be seen in Kirkeby's discussion of how power needs to be justified:

> *The leader must reconcile himself with power if he is to be worthy of our trust. But he can do this only if his power is justified by appeal to something other than himself. The legitimacy of the leader does not lie solely in the gesture of acknowledging his power, the act of admitting that he has power. It also lies in his ability to justify this power by appeal to something outside himself. The leader must be able to account for why he has chosen to seize power. And this account must always appeal to something over and above his own personal motives. Power must be chosen for the sake of something or someone. It will simply not suffice for the leader to take his power upon himself, to stand by his will to power, unless he can find another and deeper justification for his decision to operate through power.* (Kirkeby 2008:17)

Being a leader is being a leader for someone by the power of legitimacy outside the interest of the power itself. Being aware of this is what lies in the ancient Greek discussion of Phrônesis, where reaching the level of wisdom was the ambition of life. One way to sum up of a part of this is through the discussion Kirkeby (2008:170) raises in referring to the three maxims of Immanuel Kant (1929):

- The ability to think critically and autonomously;
- The ability to feel empathy for any other human being;
- Being in accord with oneself.

Being critical is a process of knowing yourself and understanding your own thinking. It is also knowing the logic in the arguments of understanding something. Empathy is grounded in goodwill and genuine caring and interest in the other. And the last maxim is to be true to oneself, but in order to do this one needs to know oneself.

Kirkeby (2008:169) points to different virtues of leadership, where a virtue is an attitude that commits because human beings have invested their whole existence in it. Virtue implies that the relation to "values" establishes commitments, and that these must have their legitimacy in a moral intersubjectivity. They must be authentic and genuine and prevent the reification of values, that is, the process of turning them into things and being measured. As seen in the discussions of the concepts of I-It. Some of virtues for the leader to strive toward could be: to be the servant of the community; autonomy; equity; practical wisdom (Phrônesis); judgment; a sense of time and the ability to seize the moment; immediate insight; articulateness; to understand organizing; and to be able to make the spirit of the organization tangible. The virtues of leadership are highly contextual, especially when it comes to the dialectical interactions in organizations. The complexity of such a context is far greater than one can completely grasp analytically because it assumes one is present and analytical at the same time, since the present continuously evolves and happens to us. Leadership as well as all organizational life are bounded and shaped by our perception and therein the ethical aspects of life. In order to develop an understanding of organizations and organizational life it is therefore necessary to analyze the role of ethics in the existential functions of humans.

References

Buber, M. 1958: "I and Thou". Continuum, London. 2. Edition. Translated by Ronald Gregor Smith.

Johnsen, E. 2002. "Managing the managerial process". Gentofte: Djøf Publishing.

Kant, I. 1929: "Critique of pure reason" (*"Kritiken der reinen Vernunft" (1781/1787)*). Macmillan, Hong Kong.

Kirkeby, F. Ole 2008: "The Virtue of Leadership". Copenhagen Business School Press. Køge, Denmark. 1. Edition 2008. ISBN 978-87-630-9950-9.

Souba, W. Wiley 2014: "The Phenomenology of Leadership". SciRes. Open Journal of Leadership, 3, 77–105. https://www.scirp.org/pdf/OJL_2014121616584278.pdf

CHAPTER 6

Existential Morality

Abstract In the current political, environmental and cultural climate, organizational ethics is not only ontologically but also societally central to organizational understanding and development. With inspiration from Løgstrup, Buber, Arendt and several others, this chapter analyzes ethics and morality as not only central to organizational culture and success but also inescapably tied to our intentional processes.

Keywords Ethics • Morality • Existential ethics and perception • The identity of corporations

Bauman concludes in his article "What Prospects of Morality in Times of Uncertainty?" that all western moral considerations stem from one of two narratives (Bauman 1998:13). The first is that morality originates from dilemmas in specific situations which an individual experiences. Ethics is a choice, albeit not necessarily one we have the knowledge or competences to accurately act in accordance with. The second narrative claims that ethics is obedience to a set of moral principles or laws for which any contradictory actions are punished. For western societies Bauman argues that both narratives have religious origins. While the notion of their origin might be correct, the exclusion of other origins seems too limiting. Like Bauman, we will for our discussions here disregard philosophies which are in line with the second narrative due to obvious inescapable religious ties.

K. M. M. Møller, M. Fast, *Investigating Being in Organizations and Leadership*, https://doi.org/10.1007/978-3-030-58138-1_6

The first narrative, however, that morality is a choice in a specific situation which depends on knowledge and competences, does seem to indicate that ethics are a matter of the capacities of human beings, and we further argue that this link should not only indicate the extent of morality but also a part of its origin. Morality and ethical considerations are experiences for us precisely because we are human beings and the complexity of our existential functions creates the ability to imagine different outcomes and different choices, and therefore we are existentially bound to navigate a sea of ethical potentials and their corresponding intentions. The same argumentation can be made regarding the origin of religion. We are religious in part because the functions of human beings provide us with such a possibility. Naturally we assume that Bauman knew this and that his claim toward the origin of morality is a comment on the history of moral philosophy. In quite the same manner our claim about the origins of morality is not intended to disregard the historical development of moral philosophy but rather to point out that the very same historical development of moral philosophy is entirely situated upon human functions and therefore has to take into account that very same existential foundation. This entails that the origin, experiences and principles of morality can potentially be extrapolated from the existential functions of humans (to the extent that our incomplete knowledge of the workings of human beings allows this). Before such a feat is attempted it is necessary to define the concepts of ethics and morality.

Morality and ethics have been defined in many different ways and an understanding will be discussed throughout this chapter, so an introductory definition is sufficient for now. Morality can be described as the aspects of existence wherein one has a responsibility toward one's own or another's existence and such a responsibility is subjected to conscious and deliberate analysis. This entails that morality can be understood as a reference to a specific branch of philosophy and as the common exercise of asking oneself: 'Were my actions justified?' Ethics on the other hand is the practical aspects of the responsibility and dialectical relationship we enter into with one another throughout our everyday lives. Because ethics is contextually dependent on nature it becomes the blurred ground where morality and everyday life interact. This distinction is necessary because it is essential to showing the complexity of ethical actions, but also because it serves as a distinction for what can be addressed in this book. Ethics is lived, messy and contextual and therefore only possible to grasp through empirical analysis. Morality on the other hand serves as the discipline of discovering through analysis the ideal principles of action. In an attempt to do so we suggest that the phenomenon of morality can be divided into at least five aspects, which are all the result of the existential functions described so far. These aspects emerge from our discussions of existential human functions so far, which include far more than the five functions we are going to suggest, and therefore we have taken the liberty of grouping the functions together based on the similarity between the principles of their functions. These aspects are *Being together, Thinking and Analyzing, Ability and Empathy, Time and Memory* and lastly *Alternatives and Self-Reflection*. We do not consider them as exhaustive discussions of ethics in everyday life, or separate entities as portrayed by their division. Rather they share the same characteristics as all attempts to analyze everyday life: they are ideal reductions and as such they are the result of our limited ability to describe them without pulling them apart when in fact they are constantly and dialectically interacting with one another.

Being Together

In an attempt to understand morality, a key component is the difference between ethics and morality: conscious analysis. It establishes a distinction between the immediate response to a responsibility and the contemplated and analyzed response, which may replace any immediate response. Because we as humans can act without conscious deliberation, we can also understand the need for that very same immediate response from someone else in some situations. This division of immersion and thinking is central to Arendt's concept of thinking (1984:14), Buber's concept if I-It and Løgstrup's (1997) concept of the ethical demand, and it is supported by the ontological discussions presented so far. As will be shown, immediate response and constant dependency are intertwined.

> *That a person is more or less in the power of another person is a fact we cannot alter; it is a fact of life. We do not deliberately choose to trust, and thereby deliver ourselves over to another. We constantly live in a state of being already delivered-either through a passing mood or in terms of something which in a fundamental way affects our entire destiny.* (Løgstrup 1997:54)

This fundamental state of dependency toward the other plays a significant role in the interpretation of the intentions of the other, so much so that Løgstrup argues that immediate action is pre-ethical and it is here that the responsibility for, or ethical demand of, someone else can be met and fulfilled (Løgstrup 1997). Thinking, conscious analysis or any similar concept as a response to the demand of another which implies a direction of attention somewhere else than toward the ethical demand posed by someone else results in a response that does not meet the ethical demand, but at the very most attempts to fulfill the social contracts we establish in our norms. Both responses, immediate or not, happen constantly and, as such, norms and values are a necessary guideline for situations where the ethical demand is unfulfilled. As Løgstrup also states in regard to law, morality and convention: “They protect us against one another, each in its own way. In their respective jurisdictions they impose limits upon the ability of one person to take advantage of another person. They prevent quite specific forms of violence” (Løgstrup 1997:54). This social safety net will be analyzed further in the following chapter.

The notion of pre-ethics is in line with the I-Thou relationship discussed previously. Both intersubjective phenomena require an immediate presence and an immediate and complete intention toward the other. Furthermore, both Løgstrup (1997) and Buber (1958), as shown in Chap. 3, argue that the disregard for this intersubjective dimension diminishes the individual’s accessibility to that very dimension moving forward. Central to the ethical demand or responsibility of the other is that most of the time the demand is unspoken and therefore rests entirely upon the empathetic ability of the individual both in regard to identifying and to accommodating the ethical demand of the other. Therefore, in regard to existence and morality there exists a state of being which is immediate, immersive and entirely directed toward the other which is central to management as well as any organizational behavior and interpretation. The fact that organizational members are people with a lifelong experience of this dimension requires that we accept the notion that all aspects of organizations are at least in part immediate, immersive and subject to our entire emotional being.

> *The greatest evildoers are those who don’t remember because they have never given thought to the matter, and, without remembrance, nothing can hold them back. For human beings, thinking of past matters means moving in the dimension of depth, striking roots and thus stabilising ourselves, so as not to be swept away by whatever may occur – the Zeitgeist or History or simple temptation. The greatest evil is not radical, it has no roots, and because it has no roots it has no limitations, it can go to unthinkable extremes and sweep over the whole world.* (Arendt 2003:95)

Thinking and Analyzing

Being together has implied that thinking stands in opposition to the ethical demand, but at the same time the quotation from Arendt states that thinking enables us to *strike roots* and *not be swept away by whatever may occur – the Zeitgeist or History or simple temptation*. This seemingly contradictory notion is central to understanding the process of morality in the individual over time. As the relationship between consciousness and unconsciousness is dialectical, so is thinking and immediate being. We can attempt to fulfill the ethical demand because of our experience, of ourselves and others, which among other things is a product of our own contemplation and the ideals and values we establish for ourselves, the notion that thinking is a process and can direct itself toward itself and therefore potentially change itself and the unconscious and immediate parts of ourselves. In order to grasp this we must, like Kant and Arendt, establish a distinction between thinking and knowing:

> *In our context and for our purposes, this distinction between knowing and thinking is crucial. If the ability to tell right from wrong should have anything to do with the ability to think, then we must be able to 'demand' its exercise in every sane person no matter how erudite or ignorant, how intelligent or stupid he may prove to be. Kant, in this respect almost alone among the philosophers, was much bothered by the common opinion that philosophy is only for the few precisely because of this opinion's moral implications. In this vein, he once remarked, "Stupidity is caused by a wicked heart," a statement which in this form is not true. Inability to think is not stupidity; it can be found in highly intelligent people, and wickedness is hardly its cause, if only because thoughtlessness as well as stupidity are much more frequent phenomena than wickedness. The trouble is precisely that no wicked heart, a relatively rare phenomenon, is necessary to cause great evil. Hence, in Kantian terms, one would need philosophy, the exercise of reason as the faculty of thought, to prevent evil.* (Arendt 1984:12–13)

Thinking as a process is central to ethical discussions, but the process of thinking rests not upon intelligence or knowledge or a wicked heart, but rather on the lack of will to perform this exercise of reason. Lack of thought and autonomous reasoning is in Arendt's perspective the main facilitator of evil. Here the concept of evil from Arendt and the concept of amoral actions from Løgstrup are considered similar, and therefore evil describes actions which lack thought and also contracts with both ethical demands and societal norms. Arendt further speculates that any process of thinking, regardless of quality or quantity, might be sufficient to prevent evil. In this regard Arendt and Løgstrup share similar considerations regarding thinking, and the application of societal norms ensures that actions are no longer evil but at the very worst hypocritical.

> *It is therefore entirely too simple to say that in the process of reflection one vacillates between doing that which is right and that which is not right. The classic characterization of reflection as a struggle between duty and inclination is therefore inadequate in as much as it fails to take into account the real character of reflection. The fact is that when the moral person does that which is evil, he or she always thinks he or she is ethically justified in doing so. Only the amoral person does evil and recognizes his or her action for what it is. In other words, the moral aspect of reflection is its hypocrisy; the possibility of one's deliberately doing wrong never really enters the mind. Thus a person who is not hypocritical is amoral. Stated differently, someone can become honest and sincere only as he or she struggles against his or her own hypocrisy.* (Løgstrup 1997:152)

The other aspect of the I-It relationship that thinking forces the individual into is, as mentioned, the ability to learn from history and imagination. It enables us to limit ourselves, for good or ill, in regard to our future actions. A central aspect of ethics is therefore our ability and capacity to process alternatives to the current moment or to reinterpret a previous or imaginary moment. This chapter and entire book for that matter are based on this very ability. If we can provide any value to the reader in regard to ethics or morality it has to come from our ability to process moments in different ways and to imagine alternative actions. But does rationality being a contextual and emotional phenomenon entail that not even our process of thinking can detach itself from its context? If so, one can argue that this book is a result of our unconscious processes as much as our conscious thinking. When this is applied to ethics a problematic notion emerges. If thinking is a process caused by several factors wherein some of them are our unconscious and immediate aspects of being, does this infer that thinking is merely another state of being wherein we are as focused and immediate toward ourselves as we are in I-Thou relationships toward the other? Are there not essential differences between the two? Our experience is certainly different, but is that contextual rather than principal? The fact that these relationships and abilities are both theoretical distinctions and states we as individuals switch between, sometimes from one second to another, does contain similarities between the direction in intentionality, which is one process carried out multiple times. Perhaps the relationships and the conscious and unconscious process of analysis are one and the very same process, just directed toward what appear to be different objects. If this line of argument is accepted this means that the process of thinking and the process of immersive being both rely on the very same principle of intentionality. Intentionality as the direction of our being, and as such evil or amoral decisions can occur only where our intentionality is not. This further enables the notion that ethics is intertwined in our perception of reality. It is a fundamental part of our thinking, immersion and intentionality. Evil or amoral actions exist only insofar as our lack of intentionality of awareness toward certain relevant phenomena allows them to exist. Once intentionality has directed itself toward something we will inevitably create a narrative wherein we are not amoral or evil, regardless of whether it is true or not. The fight against hypocrisy with the tools of critical and self-aware thinking once again becomes central to understanding and navigating both aspects of our ethical being: immersion and thinking. Analyzing something, meaning contemplating a topic or phenomenon with the application of a structured conscious approach, becomes the foundation for our conscious experience of how and why are narratives of ourselves and others unfold. If our experience of conscious deliberation or analysis is to have a place in ethics, which the nature of it representing a function of human beings ensures that it must, this is the role. So far we can conclude that the notions of thinking, analyzing, amorality, intentionality and immersion are complex and completely intertwined and dialectical. We can in other words not limit ourselves to being only coworkers or managers, because we cannot choose the experience which we use to interpret our context and ourselves. Organizational discourse and interactions must therefore conform to the societal and individual ideals of ethics and morality if they are to remain genuinely accepted phenomena. This is clearly shown in the previously mentioned notion by Mead, the generalized other, where he places the perceived attitude of others as central to the individual's ability to think and therefore to analyze contexts. The others as a part of thinking entail that concepts such as ideals, values and norms are part of the process of self-understanding and understanding of others, regardless of what the context or ourselves desire us to limit. No one is therefore simply an organizational member and every organizational member is therefore also to some extent their own private person.

Ability and Empathy

When discussing any action at least three aspects are relevant. The first is the understanding of the context wherein an action might take place, the second is the intent of that action and the third is one's ability to perform that action. All three are central to discussing morality as well as ethics as they form a crucial part of ethics: behavior. All three notions here, empathy, intent and ability, are created and limited by their interactions with themselves and the other functions of human beings. Empathy has been discussed previously and it is central to the ethical demand and its application in the *Being together* aspect of morality. Empathy and intention are concepts which are mutually dependent in regard to ethical actions; one cannot be empathetic unless one has the intent to do so, and one will not have a genuine intent directed toward the other so that the ethical demand can be fulfilled without relating empathically to the other. This does not entail that any empathetic act necessarily results in an ethical intention as the response to the empathetic act is central, as we see with hypocritical actions. An empathetic act does, however, prevent one from amoral or evil actions, as all acts cause some kind of response and therefore consideration toward the other.

The final aspect of ethical actions is the act being performed in itself. It is tempting to speak of value when discussing ethical actions as some acts surely are more desirable than others in certain situations. Actions are a result of empathy, imagination, experiences and physical capacity. It is therefore far too simplistic to attempt to describe criteria for ethical value as the actions are a product of numerous processes of which we can observe very few. The ethical value of an action is furthermore dependent on the given context and therefore we would have to move outside the parameters of morality and into an almost infinite ethical discussion (as complex as there are contexts in existence). It is, however, worth noting that given our unavoidable interpretational process we cannot prevent ourselves from judging the value of an action, nor should we, and therefore ethical value is a central discussion in any context and it is hopefully therefore a central part of the learning process an individual undertakes throughout life. As we described earlier, through our thought process and self-analysis we have the potential to change the immersive responses we give later on.

Time and Memory

The discussion of time and memory appears fairly straightforward in terms of memory as a process of recalling past experiences through the individual's current interpretational standpoint and applying the interpretation of that memory in the current situation. Time therefore can appear to merely serve as the divider between past and present. It does, however, have several important implications of organizational morality and ethics. Due to the central placement of values and social norms in the individual's thought process it is impossible to limit ourselves to only applying ethical considerations inherent in the current context. We always pull from previous contexts as well. This entails that organizational ethics has to take into account the societal norms and upbringing of the individuals as they will inevitably play a role in the future contexts of the organization. However, this is practically and theoretically impossible. Any organizational behavior cannot incorporate, nor has access to, the upbringing of each individual member, except to the extent that each member has been a part of those norms and upbringing or can through empathetic questioning and analysis develop an understanding of them. Therefore organizational ethics becomes a matter of accepting the involvement of the many different and significant contexts of society. Leaders and other culturally key members of organizations must therefore consider their positioning in regard to the dominant ethical and societal narratives used in their organizational contexts and understand it as a representation of personal attachment. This ethical relationship is central to addressing the different narratives in organizations and therefore to the internal organizational life. The relationship is also interesting and relevant for our understanding of organizations as an entity and role in societies. Dahl and Banerjee present this relationship as follows:

> *Business corporations are created and survive only as a special privilege of the state. It is absurd to regard the corporation simply as an enterprise established for the sole purpose of allowing profit-making. One has simply to ask: Why should citizens, through their government, grant special rights, powers, privileges, and protections to any firm except on the understanding that its activities*

are to fulfill their purposes? Corporations exist because we allow them to do so. (Dahl in Banerjee 2012:57)

In this description Dahl and Banerjee critique the role of business as solely aspiring to make profit. They thereby question the understanding of corporations with their reference to citizens. They create a critique based on two different concepts and corresponding assumptions, and it is clear that the concept of citizens is strongest not only in their critique but also at an ontological level. Corporations cannot exist without people, but people can exist without that specific type of corporation. We argue that this can be taken further. The notion of economic rationality and market dominance inherent in argumentations for these kinds of corporations not only excludes other discourses but also disregards the fact that profit in a closed circuit is merely relocation of resources. Profit is not solely a positive but also a specific discourse regarding the reallocation of resources from some individuals to another. Banerjee discusses the concept of economic rationality as well:

The term economic rationalism itself is problematic and needs to be unpacked. It assumes firstly that there is something inherently 'rational' about economics, which needs to be debated. Secondly, it disallows alternate imaginaries from emerging because of its discursive power to automatically label them as 'irrational'. Perhaps market fundamentalism is a more appropriate term where fundamentalism is less about the content of any belief system and more about the strength with which it is defended. (Banerjee 2012:74).

This discussion serves as an example of how understandings of certain organizations' discourses and perspectives clearly contradict societal values and norms when they are compared to other contexts. This critique is an example of how thought and time can serve as a foundation for how history and other contexts can provide us with different interpretations and critical perspectives on previously accepted phenomena wherein a lack of thought exists. This critique of profitmaking, economic rationality and market fundamentalism also serves as an example of how our ability to imagine alternatives and self-reflection is central to developing ourselves and societies.

Alternatives and Self-Reflection

The concepts of imagining alternatives and self-reflection have been thoroughly discussed already. They assume that human beings can imagine phenomena which are not immediately apparent to them, that their self-understanding consists of conscious and unconscious elements, and that both elements are interpreted and significant to our perception of contexts. This entails that our self-consciousness enables and limits the perceived options in any given context, and therefore it plays a significant role in both the fulfillment of the ethical demand and the application of social norms and values. In regard to the critique of corporations and the acceptance of social norms, one concept comes to mind which is not only in opposition to a sole focus on profitmaking but also represents the functions of memory, time and alternatives in human beings: sustainability. Anyone who attempts to grasp modern discourses and challenges will have to familiarize themselves with sustainability, and as a quick notion sustainability can easily serve as a critique of our economic, social and environmental behavior. Such a critique is essential to societies and organizations, but due to its ethical and contextual nature it is unfortunately beyond the scope of this book. Instead we argue that sustainability is an ontological concept because it is inherent in any discussion of identity, self-consciousness, self-narrative and similar introspective notions. The process of understanding oneself implies that such understanding, regardless of its precision, is identified as the same developing understanding over time. We can learn and therefore change, but it is still us, I am still me and you are still you. This implies a certain equilibrium in regard to the totality of changes amounting to no more than what our self-narrative can contain. If our self-narrative could not contain the different elements there would exist a need for different individuals. Our self-narrative is therefore sustainable, and if we wish not to impose metaphorical violence upon the other so must the ethical narratives we encounter be free of this violence. Individuals can only be considered ethically sustainable if the ethical narratives and social norms they encounter can be contained within their self-narrative. If this notion is to be taken seriously it requires that organizations adjust to the functions of human beings and disregard those discourses and ideas that are amoral or evil because they lack thought of their assumptions and consequences for the organizational members.

> *... it would be better for me to have a lyre or a chorus which I was directing in discord and out of tune, better that the mass of mankind should disagree with me and contradict me, than that I, a single individual, should be out of harmony with myself and contradict myself.* (Sokrates til Callicles, in Bowring 2011:65)

References

Arendt, Hannah. (1984) "Thinking and moral considerations. A lecture. Published by The New School". Social Research, Vol. 51, No. 1/2. http://www.jstor.org/stable/40970929. Accessed 08/04/2014.

Arendt H. 2003: "Some questions of moral philosophy". In: "Responsibility and Judgement". New York: Schocken.

Banerjee, B. Subhabrata (2012) "Corporate Social Responsibility: The Good, the Bad and the Ugly". Critical Sociology 34: 51. Sage Publications. http://crs.sagepub.com/content/34/1/51

Bauman, Z. (1998) 'What prospects of morality in times of uncertainty?', Theory, Culture & Society, 15(1), pp. 11–22. doi: https://doi.org/10.1177/026327698015001002.

Bowring F. 2011: "Comparing Bauman and Arendt. Three important differences". Sociology, 45. Sage Publications.

Buber, M. 1958: "I and Thou". Continuum, London. 2. Edition. Translated by Ronald Gregor Smith.

Løgstrup, K. E. (Knud Ejler). The Ethical Demand. Notre Dame, Indiana: University of Notre Dame Press, 1997. Print.

CHAPTER 7

A Commentary on Sustainability and Organizational Life

Abstract In the final chapter of the book we describe human beings and their functions and show how such an understanding can be the only reasonable starting point when addressing, analyzing or living within organizations. Based on this discussion we summarize the previous discussions into a coherent and ontological description of being in organizations and organizational life.

Keywords Human • Being • Organization • Dialectical • Sustainable • Intentionality • Intersubjectivity • Humanization

Through the ethical and existential discussions in this book it has been shown that any perception is always a result of emotional, historical and ethical considerations, which entails that all considerations, argumentations and scientific endeavors are in relation to a context, position and individual. By accepting this notion, any theoretical description must take into account its relation to society and relevant contexts. As such, theories and any statements cannot be said to be neutral, objective or in any other manner detached from their context and values therein, since they are a part of our perception and we are situated in our being with our biography and horizon. This chapter will therefore discuss the results of the discussions so far, along with their assumptions and consequences in regard to organizational life and organizational research.

K. M. M. Møller, M. Fast, *Investigating Being in Organizations and Leadership*, https://doi.org/10.1007/978-3-030-58138-1_7

The founding assumption of this book is that any organizational discussion must originate and be directed by our knowledge of human beings. Therefore, we have discussed human beings in regard to their existential functions and their dialectical relationship in and with everyday organizational life. We are all the result of an entangled existence whereby any careful consideration will inevitably blur the line between what is yours and mine, of me. Identity, thought and perception and what follows thereof are the result of dialectical, ethical, conscious and unconscious processes of which we merely experience the result. Despite the separation of existential functions and of everyday life made so far, it should be noted that such a division is only analytically viable. In everyday life the division is as misguiding as it is apt, since it attempts to direct attention toward a part of life we rarely if ever truly experience. Take our thoughts, our most private and uniquely individual experience: even here we are experiencing them as they have occurred, as a retrospective image upon which we attempt to infer presence. It appears as if the division between everyday life and existential functions entails a similar distance between the lived life and reflections thereof, as a psychologist might experience relating to their patients. While analytical insight provides details, it also provides a perspective which challenges the possibility of a truly immersive and empathetic experience. If there is to be one instance in which we are truly present it is one wherein we relate ourselves completely to another person. To laugh or cry with a friend or a spouse. Here we can be without self-consciousness, since the focus of our attention is not us or whether someone can benefit us, but merely the other as we are with them in the moment. Immersion, as central as it is to human connection, is therefore also a foundation for a place where we can become as close as possible to being with the other—for our puddles to become one. It is never there, but it is nonetheless the foundation for the ideal of being, of knowing and of truth, since it is there that we come as close to understanding the other as we can. A manager will find no other source of knowledge as valid and apt as immersion, but one without reflection and analytical thought is one without deliberate goals and ideals of the good.

The Complexity of the Human Being

In the discussion of organizational being the argumentation is linked to the condition that a priori the organization is the human being, and only the human being. The organization is a consequence of the human and the human is becoming a product of the organization (e.g. the discussion of dialectic, e.g. Berger and Luckmann 1966) as the interpretations and interactions with other forms, typifications and things are taken for granted. The social world is at the same time both subjective and social, as it does not appear as private but as something much more and a place where the subject is with others.

All of this is the complexity of the human being, as an ontological reflection on what the human being is in its existential functions, but also in the existential dimension of how the social can become possible, and how the subject can transcend its consciousness as it belongs to the subject and reaches into another subject's consciousness and being.

First of all, the existential functions of the subject, as discussed, are presented experiences of the consciousness and are results of emotional, historical and ethical considerations. The human being is to the world, a world confronting the subject, but the subject is also in the world. There is this fundamental dialectic of being and world that constitutes the human, and where the subject is situated with all of its being. The crux here is that the world is not in itself a force or entity that imposes itself on us. The world is and can only be *Das Ding für Uns*, as the world is only reachable by our experience and thinking of it. We cannot reach any knowing of a world in itself, as the reaching out implies a subject that acts in the world and to the world. The only one to create knowledge of the world is the human. The world does not show its selves in its existence—*Das Ding an Sich*. We can never know, and this leaves us exactly where we are, as human beings limited as human in all organizational situations.

This raises the issue of the complexity and the existential functions of being. The ontological foundation in a phenomenology of humans is that the human is consciousness and bodily movement in time and space. We are what we are in our thinking and acting, and as we are born into the world by a woman but alone, as the I is doomed to make sense of itself in this world that the I is thrown into. We develop and are equipped with our thinking and we can act, and this is fundamental to what the existential function rests upon. But this does not say anything of what we may think or how we may act in relation to and regarding what in organizational contexts and situations. This is in the becoming. There is no doubt that in relation to an existential discussion of the human being the essence of being is in the striving to understand—to understand and create meaning of oneself, of others and of the organizational world. The not understanding is the antitheses of being. We are, almost from the very beginning, conscious and orientated to the world, trying to figure it out in that we are curious about it and about ourselves in the world that confronts us. It is the center of being human to seek meaning, as it makes no sense for the I to exist in any context without trying to understand and establish some kind of meaning.

The subject is in the world as situated, and is situated with the biography and intentionality of its becoming. We are in the situation with our here and now, and with all of our memory of our selves. The situation is life and is what we are in the middle of. We are to this and to the situation with the intentionality of seeing, as we only can be aware of it from the perspective of the I, and not from a perspective placed outside the reach of the I. The I can only think what the I thinks. The thinking will be possible by the experiences of the I, but mostly of the language the I possesses. The language is the tool of thinking and we cannot think if we cannot form words and sentences about what we think. The language will be both the possibility of thinking the world and also to think the world and place the I in this experienced world, and create meanings of this world. The world will become the Lifeworld of the I. The world taken for granted, as it is known to the I in its being, is situated and bound to the world.

The transcendences of the subjects and the becoming of a social being are grounded in the thinking and the language, used to communicate and as a result of the becoming. We learn language and meaning of things in the contexts we are situated in. But all of this transcendence into the social is only possible if we are able to recognize the other as another human being. So, from our thinking the seeing of each other makes the social possible as a start. In this, the possibility to transcend and reach the other in some way is only possible if we together can create nearness. Really seeing the other and touching is grounded in the sharing of time and space. The point is how two "streams of consciousness" get in touch with each other, and how to come into understanding each other. Schutz expresses it quite simply, when he talks about the connection as the phenomenon "grow old together", to understand the inner time (*durée*) of each other. The possibility for this is that we share space and time, so we can be in a position of imagining the intentional acts of the other, when they happen. The I interprets the acts of communication in the same way as the I always interprets own lived experiences. But the I's eyes go directly through external symptoms to the internal subject of the person talking in the situation and interaction. No matter which context of meaning the I throws light on, when the I experiences these exterior indications, its validity is linked with a corresponding context of meaning in the mind of the other person. The last context must be where his present, lived experiences are constructed step by step (Schutz 1972:104). It is the I-Thou relationship as described as a context wherein individuals are truly intertwined to the point of no experience of thoughts. They are there, together, for a moment in an immediate and self-forgetting fashion. In life and in organizations there are few that we are actually close to and know in this way. On the other hand, it is clear that most of our knowing the other in the situation and in the organizational context is based upon the experience of the generalized other, as we have learned the expectations of how the others behave. In this way we are back in the circle of life where the subject learns to be social through meaningful experiences over time and in different contexts. In this we have to trust that these experiences that we take for granted are reasonable interpretations of our self in the situation and that these makes sense and work. As long as they do this, we keep on acting as we hold the reality and the truth of the meanings of it in our hand.

Organizational Life

Being is essentially dialectical, as we are constantly finding ourselves in relation to becoming immersive or becoming reflective and analytical. While there may be many existential functions, some of which we have covered insufficiently, what appears to be central for them all so far is that they are dialectical. This entails that all organizational design, understanding and life must assume this principle of dialectics. Organizations are therefore entities which are filled with assumptions and perspectives constantly interacting and being affected by one another. There is no unaffected human and therefore there can be no objective manager or leader. There is no unemotional being and therefore there can be no professional behavior or value-free ideal. And lastly there is no perception without ethics and therefore there is no neutral action, no escape toward economic rationality and its implicit lack of ethical considerations.

Within the ethical discussions so far there a notion of the good has been established. In other words, ethics is a striving toward something, the ethical demand of the other being a central figure in this. As with ethics, leadership is a striving toward the good and it happens within the perceptual parameters of ethics. Whether it is leadership by someone else or self-leadership, the direction or ideal is central to understanding the actions. Leadership is a process which affects our actions and our identity through (un)conscious input directed toward an ideal. As such, leadership is an assignment that can be undertaken by someone as a deliberate one and it is also an existential process carried out throughout our individual lives. We are, whether we want to be or not, the leader of our own life and our own organizational behavior in the same intertwined and dependent nature as those who are referred to as leaders or managers. In that regard the power a leader might possess is no more than the power given to a leader by the people they are trying to lead. The founding principle here is that the nature of the allocation and acceptance of power is central to the potential a (self-)leader has. Self-leadership therefore becomes essential to the conscious navigation of an otherwise very complex and unconscious existence through the intentional use of ideals. One might argue that the lack of allocation of power toward oneself as one's own leader hinders the very ability to navigate the sea. As such, we are existentially locked in a battle between the potential of our own leadership and the perceived objectivity of the societal structures of the Lifeworld.

Another fundamental assumption of this book is a necessary consequence of the acceptance of ethics being an unavoidable constant in life and also in science. This book is also written in an attempt to illustrate and reframe certain societal and therefore social assumptions otherwise referred to as structures, the very experience of which is contradicted by the ontological knowledge of humans. They are structures with consequences which limit and attempt to define human behavior, and because of their erroneous assumptions we argue that a revised understanding is necessary. A large part of what Critical Theory (Frankfurter School) discusses and is critical of are the dimensions in society and science that are linked to power and alienation but also to the discussion of liberation and a possible emancipation of the subject. We have previously mentioned professionalism, power, objectivity and economic rationalism, and many more aspects could and should be discussed. There remains, however, a fundamental assumption regarding organizational life and understanding. This is the understanding of what an organization is and the societal role of an organization, as shown in Banerjee's analysis of the legislature regarding corporations in the US:

> *The landmark decision of the US Supreme Court that bestowed property rights on private corporations was Dartmouth College v. Woodward in 1819. The case typified the inherent ambiguities that arise in defining the role of a corporation, ambiguities between the economic and social that are yet to be resolved today.*

> *Lawyers for Dartmouth Corporation in its move to free itself from state control argued that the rights of private corporations and private rights in general must be 'protected from the rise and fall of popular parties and the fluctuations of political opinions' (Perrow, 2002: 41). Chief Justice John Marshall concurred, declaring that 'a corporation is an artificial being, invisible, intangible. And existing only in contemplation of law' (Chief Justice John Marshall, Dartmouth College v. Woodward, 1819). Establishing the legitimacy of a 'fictitious legal person' or an 'artificial legal entity' distinct from its owners and officers (Hessen, 1979:xiv) had two effects: first, it effectively put an end to the argument that the corporation was a creature of the state thus limiting public representation and second, by conferring private rights on corporations, rights normally held by individuals, the court automatically guaranteed a system that would protect those rights. Thus, an artificial legal entity like a corporation is entitled to protection under the 14th Amendment of the US Constitution.*
> (Banerjee 2012:54–55)

The account of organizations or corporations given in the quotation above is naturally only stating something regarding the legal perception of corporations in the US and therefore the legal matter is not necessarily relevant for other countries. However, the notion of organizations as somehow separate entities transcending their employees and the corresponding societies is certainly a valid challenge everywhere in the world. In other words, our understanding of working for or in an organization has transcended into organizations being separate entities. But as discussed previously, organizations are philosophically speaking no more than a gathering of individuals with a certain communality, often referred to as a purpose, and an organization can therefore be no more than that. We cannot work for organizations, but rather we can work with other people and thereby have an experience of a phenomenon which we can then describe as an organization. Once that understanding has cemented itself, the idea that our constructed organization can somehow transcend and even equal the individuals in liberties and social rights is profoundly baffling, and more concretely it cannot be supported by any ontological argument. Ethically, the purpose of an organization, a society or any other social structure to which we grant authority must have the sole purpose of serving the common good of the corresponding society. As a leader of an organization must refer to the common good in order to find his or her legitimacy, so must organizational activities. This is not a *must* as in reference to something that only seek a moral narrowed to an activity, but a organizational *must* that is bound by ethics because human existence is. Otherwise we will continue to create entities which are in contrast to our being. This notion of social responsibility might appear to be in line with current corporate social responsibility (CSR) policies and trends, but in fact it differs quite substantially.

> *If the legal revolution that launched the modern corporation was one that served particular interests, the same can be said of the current rhetoric in corporate boardrooms about 'corporate social responsibility' and 'corporate citizenship'. The power of this rhetoric lies in its ability to validate a particular form of ideology along with its accompanying epistemological and ontological assumptions. Thus, from a critical perspective corporate social responsibility becomes an ideological movement designed to consolidate the power of large corporations.* (Banerjee 2012:53)

The analysis that Banerjee presents is a discussion of a contextual analysis of CSR in organizations. While this is certainly a relevant comment on the sincerity of previous and possibly current CSR ambitions, it is not exhaustive of the ontological status of organizations. CSR is currently a discipline carried out to meet what is perceived as the needs of the consumers. It is a subroutine in an otherwise financial machine. We argue that due to the ontological status of organizations, their main routines which are currently measured monetarily ought to be designed with explicit ambition and content directed toward the common good. While we currently work in and for organizations, from a philosophical and ethical perspective it is only logical and reasonable that organizations work for us.

The Ending: Reflections on Science and Criteria for an Organizational Understanding

A universal goal for scientific endeavors is the ideal of understanding reality as precisely as possible, based on the perceived principles of that reality. As a consequence, theories thereof which achieve an overwhelming amount of precision in their description are relatively stable over time. Throughout this book we discuss human beings in order to look at how organizations and our understanding of both have to aspire to the same ideal of precision and stability. This combined with the current environmental and political discussions as exemplified by the 17 sustainability goals presented by the UN enables the ideal of sustainable research. This is furthermore rather obvious given our argumentation for ethics as an unavoidable part of science. So far, we have discussed some social and scientific structures which are contradicted by several ontological arguments, and therefore the structures are not valid or regarded as ethical.

There are three dimensions in a discussion of this, which all could be several books. The first is from a global perspective that includes the 17 sustainable goals, a global economy and a moral discussion. It seems logical that the global challenges regarding the 17 goals are a moral discussion as they reach into people's lives in so many ways, and that this is a basic question of what the global society wants for the future. The second dimension in this is that economy is a global affair and interwoven into this discussion of the present and the future of the 17 goals. It is also clear in its logic that economy is interaction and transaction between people. We have shown this clearly in our discussion, and the argument has been presented by Kant that there is no reality in itself. It is constructed. All interaction, organization, leadership and economic is about human beings and is therefore a moral discussion. Therefore, in relation to this line of argument, this dimension of economics must also be thought as moral, as resting on a moral foundation.

The moral dimension becomes central in all that we talk of when we try to establish any discussion of a social context. Society, organizations, economy or any situation including several people can only work in a sustainable way in the light of a moral force and judgment. The problems and that which contradict morals are the forces of power and conflicts of interest. History has shown that those contradictions and conflicts are not, in any way, easy to solve or to understand.

So, in relation to a last dimension in this discussion, there is the matter of science and what science is and should be. It is from this philosophical investigation of being clear in its logic that science must have a central position. Not as the power in society judging everything, but as an autonomous and critical institution that investigates the issues we are discussing and whatever people find important. The discussion is about how and what a moral community rests upon and how science can be a part of it. Benson (1977) discusses aspects of this in relation to phenomenology and critical theory as the matter of (another) praxis, in relation to science and society, because the analysis of praxis will show which areas are important to study, and the study in itself will become a part of praxis, and a way to be part of changes. Science could in this perspective be a viewpoint on construction of alternatives. This construction is on go behind reflexivity (e.g. consciousness of consciousness—cogito cogito), and that the reconstruction of organization will be on: (a) critique of the existing discourses and settings; (b) search for alternatives in organizational life (e.g. humanization of work, development of perspectives of participations and equality, alternatives to bureaucratizes, contexts without dominance, knowledge of experts without creation of technocratic elites).

All of this is an old discussion, but the short argument for this is the moral ground and the construction of science that matters. As such, science is in no way enough, but it is important in all its areas. In this regard we hope to have shown three things throughout this book: first, how organizations can be understood; second, how science must emerge in a critical perspective that liberates and enables the individual rather than alienates and manages them; and third, that there is a necessity for discussing the other two and how being, dialectics and everyday life can be understood and related to different phenomena. The origin of all actions, all of life and science, is the human being, and as such we are entirely connected with and dependent upon each other with all the challenges and wonderful experiences that may bring.

References

Banerjee, B. Subhabrata (2012) "Corporate Social Responsibility: The Good, the Bad and the Ugly". Critical Sociology 34: 51. Sage Publications. http://crs.sagepub.com/content/34/1/51

Benson, J K: "Organizations: A dialectical view". Administrative Science Quarterly, mars, 1977, vol 22.

Berger, P L & Luckmann, T: "The social construction of reality – a treatise in the sociology of knowledge". Doubleday & Company, New York, 1966.

Schutz, A: "The phenomenology of the social world". Heinemann Educational Books, London, 1972.

Index

K. M. M. Møller, M. Fast, *Investigating Being in Organizations and Leadership*, https://doi.org/10.1007/978-3-030-58138-1

The manufacturer's authorised representative in the EU is Springer Nature Customer Service Centre GmbH, Europaplatz 3, 69115 Heidelberg, Germany. If you have any concerns regarding our products, please contact ProductSafety@springernature.com

Printed and bound by CPI Group (UK) Ltd, Croydon, CR0 4YY
30/09/2026
02225466-0002